SERVICING YOUR CAR

Marshall Cavendish · London New York Sydney

Editor: Eden Phillips
Designers: Jonathan Alden
Jane Turton

Artwork by:
Graphters, pages 60, 73
John Harwood, page 48
Nigel Osborne, pages 17, 24
Paul Williams, page 96

Printed in Great Britain by Redwood Burn Ltd
Published by
Marshall Cavendish Books Limited,
58 Old Compton Street,
London W1V 5PA
First published: 1980
© Marshall Cavendish Limited
1978, 1979, 1980
Some of the material in this book was first
published by Marshall Cavendish Limited in
the partwork *On The Road*

ISBN 0 85685 765 3

FOREWORD

The British motorist travels on average 12,000 miles (20,000km) a year. An annual check on the working parts of the car saves time and inconvenience, and makes for safer, more reliable driving. With motoring costs rising every year, doing your own maintenance makes sound financial sense – a properly tuned car runs more economically, and is far less likely to require costly emergency repairs. *Servicing Your Car* provides an easy-to-follow guide to a thorough overhaul, and helps to make breakdowns and heavy garage bills things of the past. If garage work does become necessary the information is here to assess whether a worthwhile job has been done.

The ordinary driver can quickly become an efficient amateur mechanic, using the book either as a manual for comprehensive servicing or as a reference to individual components. All the basic procedures are clearly explained, and the text is accompanied by a series of step-by-step photographs to ensure that the vehicle is serviced thoroughly. There is also a service checklist at the beginning of each chapter. This helps the reader deal with the parts of the car in a logical and methodical order. (For information on using the checklist see overleaf.) The motorist can also dip into the text to pinpoint any particular problem. There are headings at the top of each page for quick reference – 'carburettor', 'clutch', 'brakes'–and a carefully arranged and indexed list of contents.

Clear, full-colour diagrams explain how the parts of the car work, while the servicing procedures are illustrated by photographs. There is no need to relate complicated handbook diagrams to the job in hand, the actual photographs show each step to be followed. The accompanying chapters are written by experts with the do-it-yourself mechanic in mind. The text and checklist are arranged to deal in order with the top of the engine, the bottom of the engine, and then the running gear and tyres. Each chapter begins with a brief description of the relevant part of the car, going on to the order of work, preliminary checks, and the servicing procedure itself.

The book covers such essentials as safety checks on steering and tyres, overhauling spark plugs, how to maintain the air and oil filters, economical carburettor adjustment, lubrication, greasing and hydraulics. Brakes and clutch, valves and cooling systems, the ignition and electrics – and all the vital components that keep a car running – are clearly and systematically presented.

Allowances are made for variations in vehicle design, and there is a comprehensive table of engine data listing details of over 400 European, Australian and Japanese models. This provides information on cylinder layouts and tuning techniques to help the motorist get the best out of his car. Advice on where to obtain equipment and components is given at the end of the book, together with the index which makes a wealth of technical information easily accessible. Whether tinkering with the car is pure pleasure or sheer economic necessity, *Servicing Your Car* is an invaluable guide.

How to use the servicing checklist

The 16 point checklist for a thorough 12,000 mile service appears on page 6. Each of the 16 numbers refers to a chapter. At the beginning of each chapter the job in hand is clearly marked, together with the preceding and following jobs. In this way the motorist knows exactly where he is as he finds his way through the 12,000 mile service. Alternatively the checklist will show at a glance where to find information on a particular job. Simply turn to page 6 and find the relevant chapter.

CONTENTS

1 An introduction to economical servicing   6
2 Overhauling the spark plugs   12
3 Checking the contact points   18
4 Adjusting the ignition timing   25
5 Adjusting the valves   31
6 Economical mixture setting   42
7 Lubrication and greasing checklist   54
8 Checking the brakes   61
9 Replacing disc brake pads   66
10 Servicing the handbrake   74
11 Checking brake hydraulics   80
12 Servicing the clutch   90
13 Checking the cooling system   97
14 Checking the electrical systems   105
15 Checking the steering   112
16 Checking the tyres   117

An introduction to economical servicing

Changing the air, oil and petrol filters around the engine is an easy introduction to automotive engineering for the weekend mechanic—and a vital part of routine maintenance. Clean filters can produce bonuses in terms of better performance and higher mileage. Dirty filters can make performance sluggish, allow small particles of metal to find their way into vital areas and—in extreme cases—prevent a car from running at all.

Service Checklist

 1 Change engine oil. Replace oil filters. Clean petrol filters and replace if necessary.

2 Reset or replace spark-plugs.

3 Check and reset distributor points. Replace if necessary.

4 Check and adjust ignition timing.

5 Adjust valve clearances to specification.

6 Clean carburettor and adjust tick-over and mixture settings.

7 Check oil level in gearbox (manual or automatic) and rear axle and top up if necessary. Drain gearbox and axle (if directed in handbook). Attend to miscellaneous lubrication and greasing.

8 Check thickness of disc pads and brake linings. Adjust brakes where applicable.

9 Replace disc brake pads where applicable.

10 Check and adjust handbrake.

11 Examine brake flexible hoses and metal pipes for security, chafing, leaks and rusting. Top up reservoir.

12 Check clutch hydraulics for rusting, chafing and leaks. Check clutch pedal travel, and adjust clutch clearance where specified.

13 Check cooling components for leaks and blockages. Check condition and tightness of fanbelt.

14 Check operation and condition of all electrical equipment. Adjust headlamp settings if required.

15 Check all steering linkages for excess wear and movement; repair/replace as necessary. Examine driveshaft boots and steering rack gaiters for security and splits (where applicable).

16 Examine tyre wear and change wheels round if necessary. Check wheel alignment and tyre pressures.

Air and oil filters are changed about every 10,000 km or 6,000 miles as part of regular servicing. Petrol filters are usually changed twice a year—once before winter, and once before the summer holidays.

Air filter

The air filter's job is to trap dirt particles in the air supply to the carburettor. In addition, the breather pipes which disperse the slight air pressures which build up in the valve rocker box are taken to the air cleaner, where the oil-mist content is filtered out. This oil mist helps the filter to trap the atmospheric dust, but itself adds to the build-up of dirt on the filter. A worn engine is

1 The paper-type air filter is housed in a large box on top of the carburettor. To reach it, remove the wing nut and box cover

2 Dirt on the filter element shows up clearly against the light brown of the paper. A discoloured element should be discarded

3 Fit a new element in place of the old one. Replace the filter cover and align the air induction pipe according to the season

FILTERS

likely to pump extra oil mist into the rocker box.

When the filter becomes clogged, the balance of air and petrol vapour entering the engine is upset. This is unlikely to actually stop the engine, but there is a loss of power and an increase in petrol consumption. There is also likely to be an accumulation of oily deposits in the breather pipes, lowering the engine's efficiency and probably causing a build-up of sludge in the rocker box.

Hence regular checks of the air filter are essential. How frequent they should be depends on the conditions in which the car is used and on the age of the engine. A car driven regularly in heavy traffic in an industrial area, or on a dusty unsealed road, is likely to inhale more dirt than, say, one driven in a country town with sealed roads and light traffic. An old engine tends to inhale more dirt than a new one. But regular checks at, say, 5,000 km (3,000 mile) intervals will show how much dirt is being trapped and help to establish the correct checking interval.

The air cleaner is a large metal container, usually round but sometimes rectangular or tubular. It is either fitted directly on top of the carburettor or connected to it by a large-diameter pipe, and there is often an open-ended induction pipe pointing towards the exhaust pipe to breathe in air. The filter element is inside the metal container, and reaching it depends on individual design. A single thumb-screw is sometimes used to hold the top of the unit in place, or there may be a number of screws. Sometimes it is easier to disconnect the main pipe to the carburettor and rocker-box connection and lift off the entire unit.

Paper-leaves filter (figs. 1 to 3)
The commonest kind of air-cleaner element is the paper type, a concertina-like series of leaves inside a plastic mounting. The leaves allow air, but not dirt, to pass through. So the outside edges of a used filter will always be dirty to some extent.

Sometimes the contamination will be of a dry, dusty type which can be removed by tapping the filter sharply against a wall. It is more likely, though, that the dirt will be sticky, oily stuff that cannot be removed.

Usually the area nearest the air-induction pipe will accumulate most of the dirt, but the air cleaner unit is designed so that the incoming air passes right round the filter element, so that contamination on the rest of the element should be fairly even. It shows up vividly against the original colour of the element—a lightish brown or off-white.

If the original colour is obscured by black, clinging

4 The 'pancake' type of wire-mesh air filter has mesh around its edges. Another kind looks somewhat similar to the paper-type filter

5 In either type, the filter element slips out when the cover is removed. Wash it thoroughly in either petrol or paraffin

6 Oil-bath air filters are increasingly rare, but can be found in some older vehicles — like this Land Rover. Diesels also use them

7 Remove the large-diameter air hose and then the filter top. If the dirt does not just tap out rinse the filter in paraffin

8 Wipe out the bowl and refill with oil to the mark. When replacing the lid, make sure the sealing ring is in good condition

muck the filter has reached the end of its life and should be replaced.

It may be possible to rotate the element through 180° to present a reasonably clean face to the air induction pipe. But this is usually false economy: you are likely to save the price of a filter only at the expense of using more petrol.

Wire-mesh filter (figs. 4 and 5)
An alternative to the paper-element filter is the wire-mesh type, in which a form of wire wool is used to trap dirt. This type can be cleaned by swilling it in paraffin (kerosene); two or three changes of paraffin may be necessary to get rid of all the dirt.

Some older cars have air cleaners of this type which cannot be dismantled. The whole unit has to be taken off the engine and its filter end immersed in paraffin.

Oil-bath filter (figs. 6 to 8)
Although it is becoming more and more of a rarity, some cars have an oil-bath air cleaner. This type incorporates a wire-mesh filter and a quantity of engine oil (half a litre or a pint) to filter out the dust.

In this type of cleaner the wire mesh is cleaned with paraffin (kerosene). The oil is tipped out, the container cleaned, and new oil put in. Make sure that the sealing-ring gasket between the lid and the body of the unit is in perfect condition. Any breaks or distortion mean that a new gasket is needed.

Eliminating air leaks
With all types of air cleaner it is necessary, while checking or replacing the filter element, to eliminate any air leaks. Correct carburation settings depend on the efficiency of the air cleaner, and can be upset by any air that finds its way into the carburettor other than through the element.

The lid of the air-cleaner container must be fastened down tightly with no distortion caused by uneven tightening of screws. As with all other car components, groups of nuts or screws should be tightened in rotation, a little at a time, to spread the load evenly.

Pipes from the air cleaner to the carburettor and engine must be sound and free from cracks (to which they are vulnerable because of the heat in the engine compartment and inevitable oiliness around the engine). The pipes themselves should be securely tightened at their ends, and any supporting brackets also tightened.

The position of the air-induction pipe should also be checked, if it is of the swivelling type. For near- or sub-zero temperatures it should be close to the exhaust pipe, whose heat will help it vaporize the petrol. In warmer weather it should be turned away. Moving the pipe head usually entails loosening a locking-screw, moving the pipe to the correct position, and tightening the screw again.

Oil filter
Oil filters are needed because of the contamination caused by normal combustion in the engine. For each litre of petrol used, slightly more than a litre of water is formed and, while most of this is blown away by the exhaust, some remains in the engine and finds its way down into the sump to contaminate the oil. Carbon, sludge and other semi-solid impurities also find their way into the oil, some of which combine with the water traces to form more sludge.

9 Oil filters of the screw-in type sometimes stick fast because heat expands the gasket. A filter wrench is one way to move them

10 Alternatively, drive a screwdriver right through the filter body and use it as a lever. Take care not to shear off the filter base

11 Wipe clean the filter mounting on the engine block, making sure the damaged gasket has been removed — it must not be reused

12 A new gasket usually comes with the new filter; if not, buy one of the correct size. Oil the gasket lightly before reassembly

FILTERS

13 Finally, screw the new filter in place. It should be hand-tight only, but run the engine to ensure there are no leaks

14 Cartridge-type oil filters normally need to be removed from below the car. First unscrew the large cover retaining bolt

15 Still under the car, dig out the sealing ring with a screwdriver or spike. Then clean all the oil off the base of the filter housing

16 Lightly grease and fit a new sealing ring. Make sure it is level and properly seated; an oil leak here could prove highly expensive

All these contaminants can have a harmful effect. If, for example, an oilway in the crankshaft becomes blocked by sludge, a big end or main bearing could be deprived of lubricant. This could lead to the disintegration of the bearing itself and lead quickly to major damage to the crankshaft.

Two filters are provided to remove these impurities. One, which protects the oil pump itself from largish pieces of rubbish, is an integral part of the engine and is serviced only if the engine is dismantled.

The other filter deals with the finer contaminants. If it becomes choked to the point where oil cannot pass through it, an automatic bypass valve cuts in which allows the full flow of oil to be maintained. But this oil is unfiltered, and carries all the undesirable elements with it.

This filter is changed regularly, as part of the normal change of engine oil. The owner's manual for your car will say whether the filter should be replaced at every oil change, or at every other oil change. If there is any doubt, it is best to play for safety and renew the filter at every oil change. This applies especially to cars with a low annual mileage—say, 10,000 km (6,000 miles) or less—since their engines are more likely to suffer from contamination than cars used every day over longish distances.

Oil filters come in two main types: the screw-in pattern, and the replaceable element contained in a tubular shell.

Changing the oil

The best time to change oil is when the engine is at normal working temperature after a few miles drive. The warmth helps the oil to drain away completely and take the impurities with it. The oil will drain more easily still if you remove the oil filler cap on top of the engine.

Provide yourself with a container to hold the old oil (most small-car engines hold five litres—a gallon—or so), and prepare yourself for a messy job. Lying under the car, use a spanner to remove the sump drain plug. The job of changing the filter can begin while the oil is draining away.

To refill, wipe any metal 'whiskers' off the drain plug —some plugs are magnetic to help catch such rubbish— and replace the plug. Pour in enough oil to reach the 'full' mark on the dipstick. Start the motor and run it for a few minutes to allow time for the oil to fill the filter and the oilways in the cylinder block. Switch off, give the oil a couple of minutes to settle, check again with the dipstick, and top up the level if necessary.

Finally, make sure there is no oil leakage around either the filter or the drain plug. If there is, tighten them up.

Screw-in oil filters (figs. 9 to 13)

An oil filter of the screw-in type is usually about 7.5 cm (3in.) in diameter and is fitted to the side of the engine near the bottom. As a rule, removing it involves no more than grasping it firmly with one hand and unscrewing it from the engine.

But if it will not budge, first make sure you are turning it the right way—anticlockwise—and then try using an old glove for a better grip. If it still will not budge, some not-so-gentle persuasion is called for. Find a stout iron rod—a poker is ideal—and a couple of feet or so of old-fashioned clothes-line (not the plastic type) or similar thin rope. Tie one end of the rope tightly to the end of the poker, rest the poker against the side of the filter unit

17 Discard the oil filter, but retain the filter seat. This is a dished ring (centre of picture) and must be replaced right way up

18 Remove the spring and related components, noting the order in which they come out, and wash them and the filter seat in petrol

19 With the canister also cleaned and the spring back in place, replace the filter seat. Check again that it is right way up

20 Finally, insert the new filter into the canister and replace the whole unit on the car. Make sure the canister is seated squarely

and pass the line right round the filter to join the poker again. Lash it so that the line grips the filter tightly and use the poker as a lever to start the unit turning.

If this still does not work, you could hire a filter wrench, or take the whole housing off the car so you can wrestle off the filter in a bench vice.

What you must not do is to try to bang the filter round with a hammer and cold chisel, because the stub might snap, leaving you with part of the stub locked into the cylinder block.

With the old filter off, all that is needed is to remove the sealing washer, clean the face of the cylinder block where the washer was, and tightly screw in a new filter. If the old sealing washer is damaged in any way, fit a new one, lightly greased.

Cartridge-type filters (figs. 14 to 20)
Replacing a filter of the cartridge type presents no physical difficulties. The main thing to guard against is losing components as you remove the filter canister, so keep a biscuit tin or similar container handy. You will also need a new sealing ring so, if there is not one in the filter element box when you buy it, buy one separately.

A large bolt holds the canister in place. With the bolt removed, either the canister will fall away from its mounting or a little 'wiggling' may be needed. If the sealing ring between the canister and mounting does not come away by itself, pry it loose with a small screwdriver. The filter element is inside the canister and is simply shaken out into the biscuit tin.

Clean the filter bowl and other metal components in paraffin and wipe them dry. Fit the new element into the canister (some have to go in right way round), put the new sealing ring into its groove and reassemble. Do not use excessive muscle in tightening the securing bolt; wrist-tension is sufficient. But do make sure that the rim of the canister is seated squarely on the sealing ring, since an oil leak can easily develop here.

Disposable-type filters usually need about half a litre (a pint) of oil, over and above the engine's oil capacity. Hence the need to check the oil level twice when refilling the sump.

Petrol filter
The petrol filter is part of the petrol pump, which is built into the petrol supply line somewhere between the tank and the carburettor. It traps dirt from the tank before it can reach the carburettor and, hence, the engine.

Any car can suffer from the effects of a dirty petrol tank. A new car may have in its tank dirt which has accumulated during the manufacturing process. An old car will gradually accumulate specks of rust, dirt which gets in through the petrol tank filler, and even shreds of material from tank connections.

There is no hard-and-fast rule on how often the petrol filter should be cleaned because driving conditions vary so much. But a good plan is to check it once in late spring, to eliminate possible summer holiday breakdowns, and once more before winter sets in.

Cleaning the petrol filter takes only a few minutes. Finding the pump—the first time, anyway—may take a little longer. It will be either mechanical or electrical.

Mechanical pumps (figs. 21 to 23)
The mechanical pump is fitted on the engine, usually near the bottom of the crankcase but sometimes much higher—as in, for example, the Hillman Imp where it is found on top of the engine. If it is not immediately

FILTERS

21 Mechanical fuel pumps have a domed top held on by a knurled nut. To service the filter, start by removing the nut and dome

22 The filter—a metal-and-gauze disc—is removed next. Clean it in petrol, using the fingers or a brush to remove stubborn dirt

23 A considerable amount of rubbish can collect in the bowl. Clean it out and wash all components before reassembling

24 One of several designs of electric fuel pumps, which need to be removed from the car for servicing. First open the filter housing

25 Remove the cartridge-shaped filter, and clean it and the filter housing thoroughly by brushing them with petrol

26 When reassembling, check that all sealing rings and gaskets are in immaculate order. Replace with new any faulty ones

recognizable, the failproof way of finding it is to trace the feed-pipe back from the carburettor to the pump which, because it is driven by the engine, must be in its immediate vicinity.

Mechanical pumps usually have a slightly domed cover, held in place by three or four screws. Removing these will allow the dome to be lifted off the pump body, exposing the filter.

The filter is a circular metal-and-gauze wafer, and the dirt will be visible on the gauze. The worst of the muck can be removed—carefully, to avoid damaging the gauze—with a piece of card, a brush or even one's finger. A quick swill with petrol will shift the rest.

Petrol will tend to dribble out of the pump as the filter is being checked, and this is a fire hazard. Avoid the temptation to use a match or cigarette lighter, even if working on the roadside in a winter gloom.

Be careful, too, not to snag your hands on the retaining screws.

Electric pumps (figs. 24 to 26)

Electric pumps are more of a problem, since they are independent of the engine and can be mounted almost anywhere. The make of car is unfortunately no guide. On the Morris Minor, for example, the pump was ideally placed on the bulkhead near the battery, where it could be instantly reached; on the Mini, it is almost inaccessible beside the rear-wheel arch. The depths of the boot is another favourite hiding place, often made more obscure by some form of shield over the pump.

However, electric pumps give off a rattling noise when the ignition is turned on so can sometimes be found by ear. If not, the only way is to trace the fuel line from the carburettor towards the tank until you see a cylindrical object with an electric wire leading into it.

An electric pump has two pipes coming from it, one from the tank and the other to the carburettor, and the filter is usually between these pipes. Before opening the pump, switch off the ignition. The dirt might stop the pump working, since the fuel flow is interrupted, but removing the filter with the ignition on will make the pump start up again and spew petrol everywhere. Quite apart from the dangers of naked lights, a spark from the pump could trigger off an explosion. Also, have some kind of clamp or plug ready—the end of a plastic bag and a rubber band will do—to stop petrol dribbling out of the inlet pipe.

Unscrew the pump with a spanner, and clean the filter (which can be a gauze tube or gauze disc) by brushing it lightly with petrol before replacing it.

Line filter

As an alternative to—or sometimes to supplement—the filter in the fuel pump, a line filter is used in some cars. This takes the form either of a plastic disc about 7.5 cm (3in.) in diameter, or of a length of plastic tube of a wider gauge than the fuel line. It is inserted into the fuel line near the carburettor. Usually a line filter cannot be cleaned, but has to be replaced. But it is just a few minutes' work to remove the Jubilee clip at each end of the filter, remove the filter and discard it, and fit a new one. The replacement does not even have to be identical to the old, so long as its diameter fits your fuel line.

As well as the filters in the fuel line itself, some cars have a small extra filter inside the carburettor just where the fuel line enters it. Individual makes of carburettor, and adjustment procedures, are dealt with in chapter 6, *Economical mixture setting*, on page 42.

Overhauling the spark plugs

Cleaning and 'gapping' spark-plugs during the service routine, say every 5000km or 3000 miles is a relatively simple job. So simple, in fact, that you may be tempted not to bother – especially if the engine is running well and starts easily.

But this happy state of affairs cannot continue indefinitely. All the time you are driving, the gap between the electrodes of the spark-plugs is increasing by the small, but significant, amount of 0.025 mm (one thousandth of an inch) every 1,600 km or 1,000 miles.

Eventually, as higher and higher voltages are required to jump the gap, you will find the car difficult—or, in cold weather, impossible—to start. Engine running may become ragged, and fuel consumption will increase.

This is why overhauling the spark plugs is an essential part of the 12,000 mile service.

Spark-plug basics
A spark-plug's function is to provide a small air gap in the high-tension electrical circuit of a car. In trying to cross this gap, the HT current creates a spark, and the spark ignites the fuel vapour in one of the engine's combustion chambers.

Although highly volatile in air, the vapour needs a 'fat' spark to ignite it when compressed. To generate this spark, a comparatively wide air gap is necessary. And to bridge the gap a high voltage—normally about 8,000 volts—is needed. This voltage is provided by the car's coil, and relayed to the spark-plug through the distributor.

At an engine speed of 4,000 revolutions per minute (about 110 km/h or 70mph in an average car) each plug in a four-stroke engine gives 2,000 sparks a minute, and each single one of these sparks has to occur at precisely the right time—a split instant before the piston reaches top dead centre on the compression stroke.

In the combustion chamber of the engine, the temperature of the insulator nose must stay at or below 850°C, or it will become so hot as to ignite the vapour before the pre-set time. (This condition, pre-ignition, causes acute misfiring, loss of performance, a loud knocking noise and a high degree of engine wear.)

On the other hand, the operating temperature of a plug should remain about 550°C or carbon deposits will rapidly build up. And if the temperature should fall below 350°C, the plug will start to oil up at low engine speeds.

When to change plugs
In spite of the stresses on them, modern spark-plugs have an average life of 12,000-16,000 km (8,000-10,000 miles) in four-stroke engines. It is possible to keep plugs longer, but a plug that is nearly worn out can result in uneven running and a 20 per cent increase in petrol consumption, and can make the car difficult to start.

Further, an unreliable or neglected plug can be a safety hazard—for example, should it cause the engine to hesitate or misfire while the driver is overtaking. So to be on the safe side it is best to change the plugs for new ones every 16,000 km and to clean and re-set the gap every 4,000/5,000 km.

To get the maximum benefit from a set of plugs, start with a brand new set, a new set of contact-breaker points for the distributor (pp 18-23) and new plug leads if the old ones are worn. All three items are part of the spark-producing 'chain', one weak link in which can undo the benefits of the other two.

Plug leads should be replaced if they have obvious physical defects like cracks in the insulation or sections where excessive heat has defaced it.

Buying spark-plugs
It is best to shop around when buying plugs. During a garage service you usually pay the suggested retail price for these items, but you can get a substantial discount by buying at an accessory store or taking advantage of special offers.

If in doubt which type of spark-plug will best suit your car, the simplest solution is to stay with the make and grade of plug used as original equipment. (As car handbooks are occasionally out-of-date even when printed, check with a reputable dealer if you have any doubt.) As the dimensions of a spark-plug can vary quite considerably, it is better not to guess which size to buy; instead check this point too with your dealer.

A frequent cause of plug trouble is town driving, where the engine is idling a great deal and speeds up over short distances only. In this situation, a layer forms over the plug electrodes and acts as a form of insulator. Once this layer reaches a certain depth it prevents full efficiency at wider throttle openings, resulting in mediocre acceleration and a lowering of maximum speeds.

In this case, sandblasting away the layer (see p 14) will erode an excessive amount of metal from the plug. One way to 'clear' the plugs is gradually to build up speeds to disperse the layer. A short, progressive burst of acceleration often has the same effect.

But if these problems happen every time your car is driven in town you are probably using the wrong grade of plug. To decide which plug grade to change to, seek the advice of a reputable dealer.

Removing the leads
The first stage in servicing spark-plugs is to remove their HT leads. But first you need to number them (fig. 1) since, if you connect the wrong lead to any plug, the engine could fire in the wrong order and perform very badly indeed.

For this job, it does not matter which order you use in numbering the leads—as long as you remember it.

Service Checklist

1 Change engine oil. Replace oil filters. Clean petrol filters and replace if necessary.

 2 Reset or replace spark-plugs.

3 Check and reset distributor points. Replace if necessary.

PLUGS

1 To make sure that you return each of the HT leads to the right spark-plug, number each lead with a piece of adhesive tape

2 Before you remove the spark-plugs, use a soft brush to clean off any dirt or grease in the area around the spark-plug holes

But for some other tasks about the car, you will need to know which cylinder is No. 1, which No. 2 and so on. So if the correct order is given in your owner's manual, use it now to save confusion later. (See also pp 35-41).

Always remove the plug caps by pulling the caps themselves—not the leads. The caps are not difficult to remove, and using the leads to tug them off could damage the leads enough to interfere with the flow of current.

Be very careful about touching the plug leads or pulling off a plug cap when the engine is running. The high voltage running through the leads can give you a nasty jolt and, when a plug cap is removed, the voltage tries to jump the gap and rises in consequence.

The 'bite' from a plug lead is not dangerous in itself because, even though the voltage is high, the current is correspondingly low, so there is no danger of electrocution. But whipping your hand violently away from the source of the shock could send it into the fan or onto the hot exhaust manifold.

Removing the plugs

Before removing the plugs, <u>clean off any dirt</u> that has accumulated around them with a soft brush. To remove the plugs it may be necessary to buy a special plug spanner (fig. 5). Some plugs are more difficult to remove than others, being deeply recessed into the cylinder

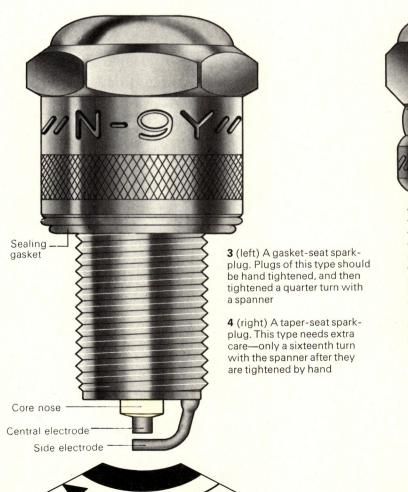

3 (left) A gasket-seat spark-plug. Plugs of this type should be hand tightened, and then tightened a quarter turn with a spanner

4 (right) A taper-seat spark-plug. This type needs extra care—only a sixteenth turn with the spanner after they are tightened by hand

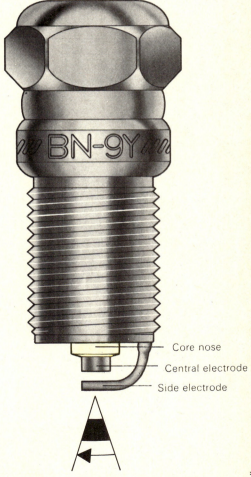

5 Several types of spark-plug spanner are available. Make sure you choose a spanner that can reach all the spark-plugs

6 Rest the spanner squarely over the spark-plug. Do not tilt the spanner, as this could easily crack the plug's ceramic insulation

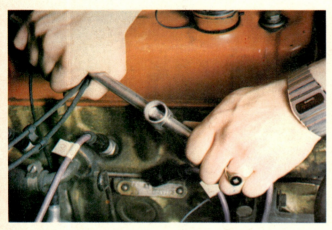
7 You should not have to apply much force to undo a spark-plug. If a plug is hard to move it has probably been over-tightened

8 With the spark-plugs removed, clean their electrodes in white spirit. Alternatively, have the plugs sand-blasted by a garage

head, while others are set at odd angles. In most cases, though, a box spanner long enough to clear the top of the engine—plus a lever to provide extra turning force—will do.

Whatever tool you use, make sure it remains dead square to the plug. A spanner tilted to one side could crack the plug's tough ceramic insulation, making the plug useless.

Start by sliding the lever through the top of the box spanner so that there are equal lengths on either side for a two-handed grip (fig. 7). Leaving the whole length coming from one side only will give extra leverage, but will also tend to tilt the spanner.

If the plug proves hard to turn, try using your thumbs against the box spanner. Alternatively, try a longer lever, this time with most of its length protruding from one side of the spanner. Use one hand to keep the spanner square while the other applies force to the lever.

Cleaning the plugs

Once all the plugs are out, examine the plugs carefully —their colour and condition can tell you quite a bit about the condition of your engine. Ideally, the plug 'points' (the correct name is electrodes) and insulator nose should be a milk chocolate shade with a thin coating of carbon dust over them. A grey-white colour indicates an over-weak mixture, while a layer of black, fluffy soot is a sure sign of an over-rich mixture. This can be corrected quite easily when adjusting the carburettor settings. (See chapter 6).

Clean the spark-plugs with white spirit and a rag. If this does not get the electrodes clean, you will probably need new plugs. You could try taking the old ones to a garage to have them sand-blasted—a thorough process which removes all the heavier carbon, dirt and oil deposits—but you should get a price for the job before going ahead. Some garages charge as much for half an hour's work as you would pay for new plugs.

Do not use a wire brush on the electrodes. Bits of the wire will adhere to the ceramic of the plug and eventually the power will 'track' down the surface of the plug instead of creating a spark.

Re-setting the electrode gap

To re-set the gap between the electrodes you will need a feeler gauge (fig. 9). Feeler gauges come in both metric and inch sizes—for example, 0.6 mm or 0.025in. (stamped '25' to indicate 25 thousandths of an inch). Before buying, check in your owner's manual what the correct spark-plug gap should be.

To check the gap, slide the feeler gauge between the electrodes and make sure it is a tight fit. If the gap is too wide, tap the side electrode gently with a screwdriver handle or similar light tool. If it is too narrow, use a knife slipped under the side electrode and carefully lever it upwards a fraction.

After 10,000 km (6,000 miles) it is likely that the underneath of the side electrode will have become eroded to a noticeable extent. It is important to have a flat face opposite the central electrode and to achieve this, a points file (a cheap tool available from motor accessory stores) is needed. The electrode should be bent carefully upwards, using a broad-bladed knife or purpose-made gapping tool, until the file can be used. File the face

PLUGS

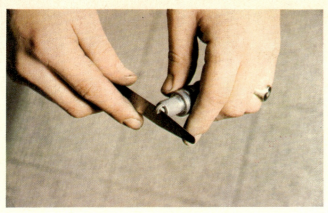

9 The electrode gap must be accurate to give a good spark. Use a feeler gauge to set it to the figure given in your handbook

10 As an alternative, this small tool from Champion combines a feeler gauge with a small device for bending back electrodes

11 If the electrodes need to be filed smooth, bend the side one with a broad-bladed knife or with the Champion tool

12 File the face of the central electrode with a points file (from accessory shops). Do the same to the other electrode

evenly until it is flat again and bend the electrode back until the clearance is correct.

If the plugs are badly pitted and corroded after this mileage, replace them.

Replacing the plugs

One reason for plugs being hard to remove is that they have been put in too tightly in the first place, or that some obstruction may have found its way into the threads of the plug or plug hole. Putting a few drops of engine oil on the thread of each plug is a fairly common practice, but do not use graphite instead of oil—it makes the problem worse.

The main thing to remember, in replacing plugs, however, is not to over-tighten. Hand screw the new or cleaned plug into place then, if it is a gasket-seat plug (fig. 3) give it a quarter turn with the spanner; if it is a taper-seat plug (fig. 4) give it a sixteenth turn. If you have a torque wrench, the maximum setting for tightening gasket-seat plugs is 25lb/ft, and for taper-seat plugs it is 8lb/ft.

Checking the coil connections

One matter that is easily overlooked is making sure that the LT wires to the coil are connected to the correct terminals. Reversing the coil connections can reduce the voltage by a massive 30 per cent.

On the coil is a heavy lead going to the distributor cap. There are also two thinner leads, connected to the + and − symbols on the coil. One of these wires goes to the side of the distributor and the other to the ignition switch, but which goes where depends on the car.

On a negative-earthed car, the wire from the distributor must be connected to the − symbol on the coil. On a positive earthed car, the distributor wire goes to the + terminal on the coil. If they have been reversed, you simply swap them around at the coil.

On older cars, the coil may be marked SW and CB, and you need not check the car's polarity. The CB wire goes to the distributor (contact breaker) and the SW wire to the switch.

Checking the HT leads

Spark-plug leads are subjected to heavy stresses from heat, oil and mechanical effects. For high-tension electricity of around 10,000 volts to flow back to earth via the air gap of the spark-plug calls for an effort—and if faults in the insulation of the lead allow an easier path to be taken then the electricity will accept the opportunity. Sparks jumping from the lead to the engine are a good example of this.

A temporary cure is to spray the lead in question with a silicone-based de-watering fluid from an aerosol, such as WD 40. Although this will—literally—fill the gaps, the only proper cure is to fit new leads. If one lead is giving trouble, it is reasonable to assume that the others will soon be doing the same.

Changing the leads

New leads come in two types. You can buy the ready-fitted sort, cut to length and fitted with terminals. Or you can make up your own from a length of cable.

Whichever you buy, it is important to have the correct type of lead. Carbon-cored leads, for example, are

15

13 Make sure that the LT wires to the coil are connected to the correct terminal. These coil terminals are marked + and −

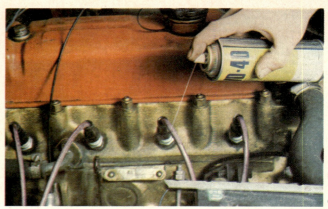

14 If the insulation of the HT leads is faulty, you can fill the gaps temporarily by spraying the leads with a fluid like WD 40

15 These HT leads are located in the distributor cap by pointed screws. You will need a slim screwdriver to remove the leads

16 On other distributor caps, the HT leads are held in place by knurled plastic thimbles. Undo the thimbles to pull out the leads

designed for radio interference suppression; copper-cored ones, by themselves are not. If in doubt, a dealer handling your make of car will advise you on the most suitable type of lead.

The length of the new leads should match that of the old ones. Too short a lead will result in acute bends, throwing an extra mechanical load on the leads especially when the engine rocks in its frame at low speeds. Over-long leads are untidy and may touch the hotter parts of the engine.

Do not be tempted to tape or bind plug leads together, as this may result in 'crossfire' from one lead to another. Crossfire occurs when one plug lead's close proximity to another causes its plug to fire out of sequence and possibly at the same time as another plug.

At the plug-cap end, fitting is usually made by screwing the cap on to the lead. This simple arrangement gives a high degree of mechanical support besides making a good electrical connection. Chipped or cracked plug caps should be renewed to avoid the possibility of voltage-leak (or 'tracking' as it is usually known). Wipe off any dirt with a rag.

Testing the security of the connection by pulling the lead is not a good idea—the internal electrical connection could be strained. If the cap is screwed fully home, the connection must be sound.

Connecting to the distributor

At the distributor end, the plug leads are usually held by pointed screws reached from inside the cap (fig. 15). A slim screwdriver is needed to remove the old leads and fit the replacements. Be careful here, for a screwdriver whose blade is a tight fit inside the screw-holders might crack the plastic, thus ruining the cap.

As the old leads are removed, clean the insides of the distributor cap sockets with a twisted-up piece of clean, dry rag. Then push the new lead inside the socket as far as it will go, and tighten the screw. It will take a certain amount of effort to pierce the plastic insulation but there will be no doubt when it is fully home. Do not over-tighten the screws; ordinary wrist pressure is enough.

Other types of distributor have knurled plastic thimbles on top of the lead sockets which hold the leads in place. With these, you undo the thimble, remove the copper washer and slide it off the old lead (fig. 16). The end of the replacement lead should have about 3 mm ($\frac{3}{8}$ in.) of the insulation carefully cut round with a sharp knife, taking care not to cut through the conductor wire.

Carefully unravel the strands of the conductor wire, and then spread them out and bend them back on themselves to cover the cut end of the wire and copper washer. They will then make firm contact with a metal insert in the socket in the distributor cap, and the contact will be maintained by the physical pressure of the plastic thimble onto the copper washer.

Before pushing the end of the lead home, look inside the socket to make sure that the metal is free from corrosion. Any green-coloured deposits or black burning traces can be easily removed by careful scraping with a screwdriver tip or a piece of sandpaper.

Finally, refit the leads in the position in which the manufacturer originally fitted them. This avoids the risk of crossfire between the leads.

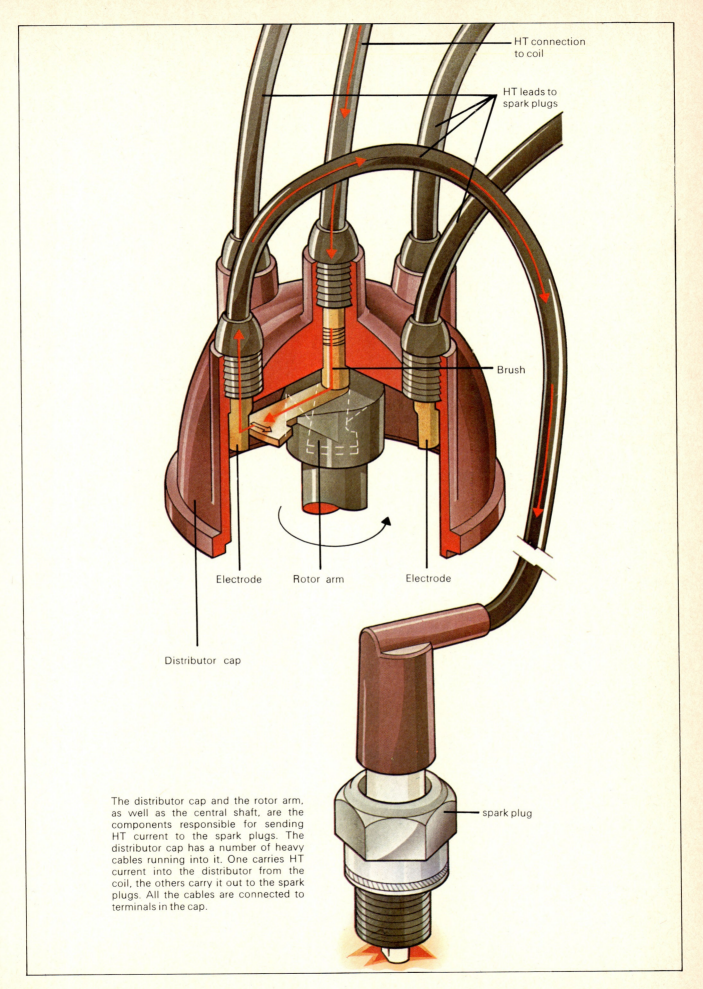

Checking the contact points

Incorrectly set contact points cause the engine to operate below its optimum performance. The resulting increase in running costs is far in excess of the cost of re-adjusting or replacing them.

Checking and adjusting or, if necessary, replacing the points is a straightforward job with which the d-i-y mechanic can easily cope. Normally, the procedure is to check the points gap every 10,000 km (6,000 miles) and to replace them when they have done between 16,000 and 20,000 km (10,000 and 12,000 miles).

Checking the distributor cap
Before you deal with the points, start by looking inside the

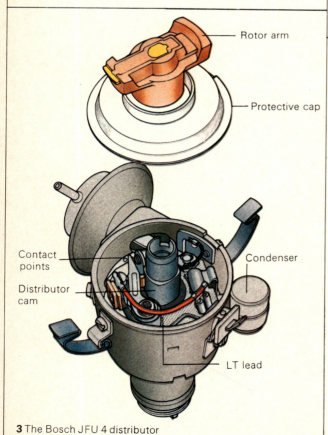

1 A Hitachi distributor

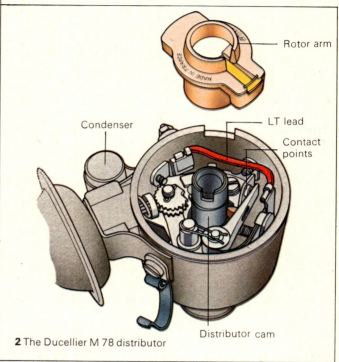

2 The Ducellier M 78 distributor

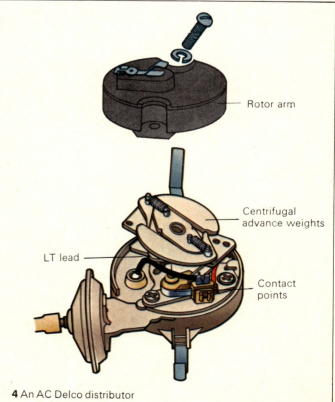

3 The Bosch JFU 4 distributor

4 An AC Delco distributor

POINTS

Service Checklist

2 Reset or replace spark-plugs.

 3 Check and reset distributor points. Replace if necessary.

4 Check and adjust ignition timing.

distributor cap. With the ignition switched off, unclip the cap, wipe both the inside and the outside with a clean rag and then examine the cap closely.

If there are any hair-line cracks in the cap, or any pieces broken from the base or outlet sockets, you will need to renew the cap. Any mechanical fault of this nature tends to encourage damp and dirt to build up and these deposits provide a path to earth for high-tension electricity. In dry, warm weather such faults may not be noticed—but they will be in misty, cold weather when nearly all the battery's output is used to turn over a stiff engine.

On top of the cap you will find an HT lead to each of your spark-plugs. These must be attached to the correct terminals on the cap, or your cylinders will fire in the wrong order—an unnerving experience. So when fitting a new cap, transfer the leads one at a time from the old cap to the new, working clockwise around both caps. Alternatively, if you have to remove the old cap and take it to an accessory store while you find a replacement, number all the leads in order.

Checking the carbon contact

The carbon contact in the middle of the cap is the connection from the coil. It must be in sound shape to carry electric current from the coil to the distributor and, ultimately, the spark-plugs.

On Lucas distributors, the contact is a spring-loaded carbon cube or 'brush'. To work properly, it must be near enough square-ended and, at rest, at least 3 mm ($\frac{1}{8}$in.) must be protruding. If the contact does not satisfy these conditions, or if the tip is cracked or broken, pull the contact and its spring out of their socket and fit a new one.

On other types of distributor, the contact is a fixed one and there is a spring tongue on top of the rotor arm which bears against the contact. It is unlikely that there will be much wear on a contact of this type but it should be renewed if broken.

Cleaning the rotor arm

Every distributor has a rotor arm on top of its spindle. This is a hook-shaped device with a metal face which almost—but not quite—touches the four, six or eight contacts in the distributor cap which are connected to the plug leads. The current from the coil comes through the central connection

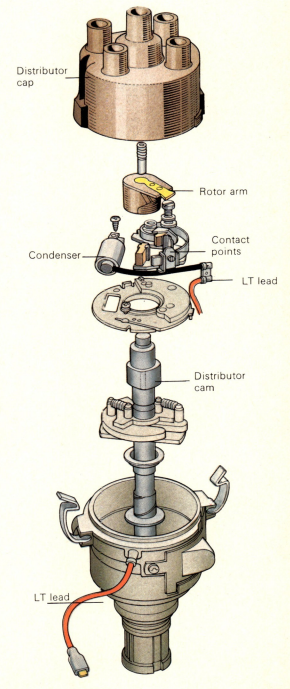

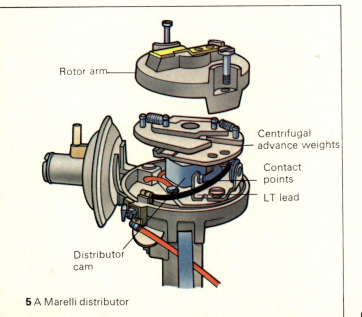

5 A Marelli distributor

6 The Lucas model 43D distributor

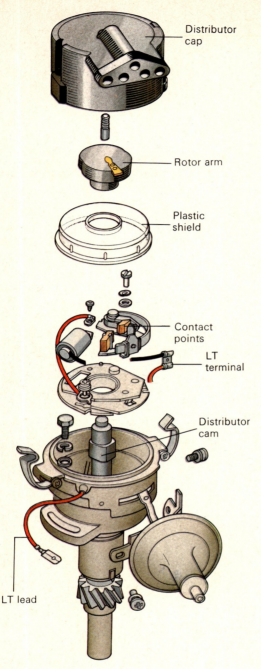

7 The Lucas (early) model 44D distributor

it is a tight fit and will not easily come free. In this case, gently lever the rotor arm upwards from its base with a broad-based screwdriver pivoted on part of the distributor body. Go cautiously—the distributor is easily broken.

The contact-breaker points

There are two contact-breaker points, both metal. One is fixed (but adjustable) and earthed through the body of the distributor. The other is spring-loaded and is moved by an arm operated by the distributor cam (fig. 7).

The contact-breaker points are, in effect, an automatic switch to make and break the electrical circuit which indirectly feeds HT current to the spark-plugs.

In a four-cylinder engine running at 4,000rpm, the points open and close 8,000 times a minute. Since they have to open and close completely each time, the gap between them when they are open must be exactly right.

The gap between the points should be checked at 10,000 km (6,000 mile) intervals. Since the size of the gap varies from car to car you will need to consult your owner's handbook to find the recommended clearance.

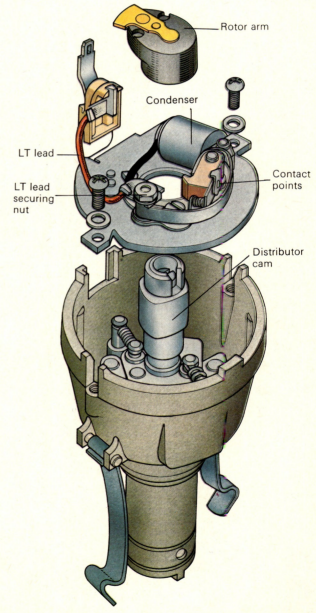

8 The Lucas model 23D distributor

in the distributor cap, runs along the metal strip on top of the rotor arm, and jumps from the rotor arm face to each plug lead contact in turn.

There is a tiny air gap between the rotor arm face and the contact, which means that a certain amount of sparking is inevitable as the rotor arm passes each contact. This sparking will result, after some time, in pitting and discolouration of the metal on the rotor arm. Discolouration can usually be removed with a rag or by rubbing the rotor arm against a tyre, but pitting is best dealt with by fitting a new rotor arm.

It is possible to get rid of pitting by filing or sanding the metal clean. But this will increase the air gap, increasing the voltage demands of the ignition system and, possibly, causing difficult starting in winter. So, if there is any doubt about the condition of the rotor arm, it is best to replace it.

Although you can usually remove a rotor arm easily by pulling it straight up off the spindle, there are times when

POINTS

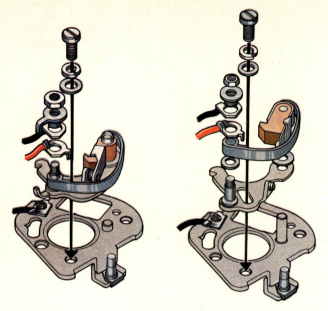

Checking the gap
The points gap is checked in exactly the same way as the gap in a spark plug – by sliding a feeler gauge (page 14) between the contacts.

First, however, you must switch off the ignition and turn over the engine by hand until the points are fully open. There are several ways of doing this:

Use the starting handle (if your car has one)

Engage top gear, release the handbrake and rock the car backwards and forwards

If you already have the spark-plugs out, thereby reducing the compression in the engine, pull on the fanbelt (not the fan blades)

Use a spanner on the crankshaft pulley

The arm which operates the movable point rests against the lobe on the distributor cam in such a way that the points faces separate and reach their maximum opening when the lobe is at its peak (fig. 18). This is the position in which the points gap is checked.

When the points are fully open, it is likely that you will see a small build-up of carbon and possibly a slight 'pip' between the two surfaces. These must be removed before checking the gap.

9 Two typical contact sets: (left) a one-piece contact set, and (right) a two-piece set

10 Two one-piece contact sets. They are of different manufacture and, because of their shape, are not interchangeable

11 To change the points, first remove the distributor cap (with the ignition off). Examine the cap carefully for signs of wear

12 Remove the rotor arm. It may pull away easily, or it may need gentle leverage. At all times avoid straining the distributor

13 Undo the screw or nut connecting the LT and condenser leads to the points. Make a note of the order in which they are fitted

14 Pull the leads clear of the base plate on which the points are mounted, and make sure they are not placed under any strain

15 Two screws are usually used to secure the points. One mounts in a slot to adjust them. Remove the other screw first

16 The distributor cam and points (closed)

17 Points open at too large a gap

18 Points open at the correct gap

19 Undo the second screw and place both in a safe place ready for reassembly. Be careful not to drop them into the distributor body

20 The old points can now be lifted from the base plate. If they are not badly worn, they may be cleaned as an emergency standby

With the points still fully open, gently stroke a points file (page 14) across the peak of the pip until it has been eased flat.

To remove the carbon deposits, first turn the engine over again so that the points are closed. Then pull back the moving point with your fingertips, and insert a small strip of 400 grade wet-and-dry abrasive paper between the two points. Let the points close up again and move the paper

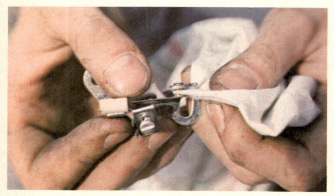

21 Before fitting the new points, the two faces must be carefully cleaned with petrol to remove every trace of protective coating

22 Having wiped the base plate clean, fit the new points in place with the screws. Do not fully tighten the two screws at this stage

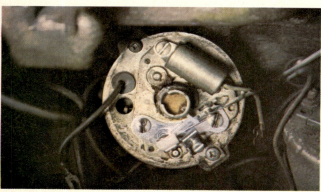

23 Turn the engine over by hand until the distributor cam opens the points. Then insert a feeler gauge of the correct thickness

24 If the points gap is incorrect, it can be adjusted by levering with a screwdriver in a small cut-out at the base of the points

25 Recheck the gap with the feeler gauge and, when it is correct, with the feeler still in position, gently tighten the two screws

26 Replace the distributor cap. It can only be located in one way, since a raised square on the cap locates in a cut-out

POINTS

backwards and forwards until all the carbon has been removed.

Once the points have been cleaned up, turn the engine until they are fully open again. Then slide the correct feeler gauge between the two contacts. Ideally, it should be a reasonably tight fit, but not so tight that you have to force it.

If the gap is incorrect, you can enlarge or reduce it by loosening the locking screw which holds the fixed point in position and moving the supporting base plate in the appropriate direction (fig. 25). Finally, tighten the locking screw firmly and re-check the gap.

If the carbon build-up on the points is such that you cannot remove it, or the points themselves are badly pitted, they are due for replacement.

Replacing the points

Contact-breaker sets are available either as the one-piece type (fig. 9) or as the older, two-piece type (fig. 9). The one-piece type makes the job of replacing points a great deal easier and, on many cars, it is now possible to swap two-piece contact-breaker sets for the one-piece type, but not generally vice versa. Motor accessory stores usually sell both types.

When buying replacement points, whether one-piece or two-piece, take with you a note of the year, make and model of your car together with the type (and if possible, the serial number) of the distributor.

Before you remove the existing points, compare them with the new set that you have just bought. Unless, of course, you are changing from two-piece to one-piece the new and old points should match exactly.

One-piece contact sets

Changing a one-piece points set is simple. All you have to do is to remove the rotor arm and disconnect the wire from the points to the distributor body. The locking screws are removed and the set is then free to be lifted up and off.

Before fitting the replacement set, wipe the faces of both points with a rag moistened with petrol to remove any protective coating and make sure that the underside of the base plate is free from any similar coating. Finally, tighten down the new set, check the gap and then firmly tighten the locking screw.

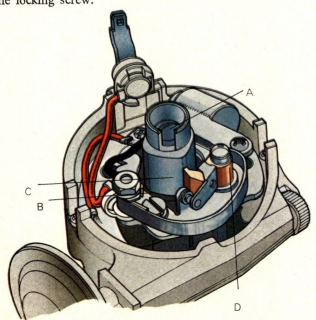

27 For routine maintenance, lightly oil points A and B with clean engine oil, and smear points C and D with a thin coat of grease

Two-piece contact sets

There are several different makes of distributor, such as Lucas, Bosch, AC Delco and Ducellier, but the routine for fitting a two-piece contact set is substantially the same for all of them. Two things will help you to avoid mistakes:

First, when dismantling the distributor, write down how and in which order the various components are removed. Although some contact sets come with an instruction sheet, it is essential to write down the positions of the various washers and insulation items to make sure that they go back in the same way.

Second, make sure that the points assembly—and in particular, the moving point—is fully insulated. You can do this by seeing that all insulation components are in the right place and—most important—right way up, and by looking carefully at the finished job. Failing to do so will result in a short-circuit of such proportions as to cause the wiring to smoulder and char in a very short time.

Generally, to change a two-piece contact set you need to remove the rotor arm and disconnect from the points the wires coming from the LT side of the ignition and the condenser. Then you loosen the retaining screw(s) and carefully lift out the points.

Next, as with one-piece contact sets, you wipe the preservative from the new points with a petrol-moistened rag, and reassemble in reverse order. Finally, you check the gap and tighten the locking screw.

An example of a two-piece contact set is illustrated in fig. 9.

Lubricating the distributor

When replacing or checking the points, it pays to lightly lubricate the distributor. Give the faces of the cam a thin smear of oil or grease (for example, 3 in 1 oil). Then apply a few drops of oil through the gaps in the distributor base plate, and finally lightly oil the pivot post.

Do not over-lubricate the distributor, however. If any grease or oil finds its way between the points, excessive carbon deposits will occur and your car will misfire badly, reducing performance, and perhaps leaving you with the job of having to strip the distributor to clean it up.

Replacing the condenser

The condenser (fig. 6) has two functions. One is to store an electrical charge and the other is to reduce the amount of sparking between the contacts as the electrical flow is interrupted.

Although the spark at the points is a tiny one and the burning effect of each operation can hardly be measured, the burning accumulates over the hundreds of thousands of on-off operations. The condenser's job is to minimize this.

If the points show signs of excessive burning and there is a corresponding roughness or hesitation in the running of the engine, the condenser is probably at fault.

Leaving a faulty condenser will eventually result in the points becoming so burned that soon they will not work at all. A new condenser is inexpensive to buy, so if there is any doubt about its condition you should renew it. The job is straightforward: you disconnect the wire from the distributor, remove the retaining screw, fit the replacement and replace the screw.

Finally, remember that any ignition system—no matter how well maintained—is susceptible to 'tracking', especially in damp weather. A monthly wipe inside the distributor cap and around the plug leads and caps to remove dirt and moisture takes only a few minutes and is well worth the trouble.

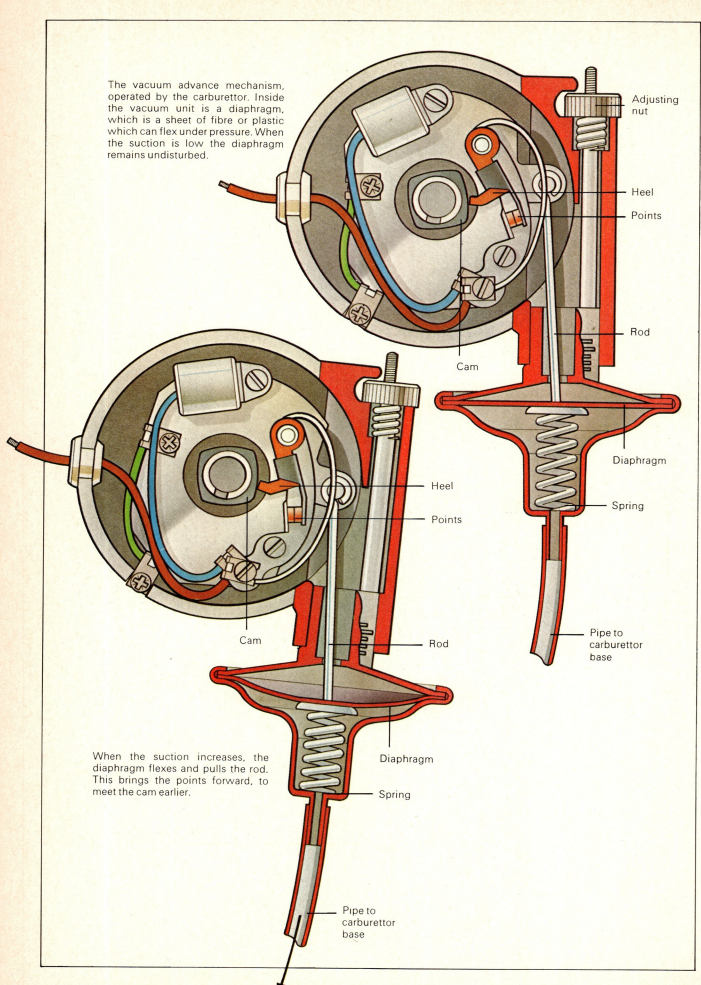

IGNITION

Adjusting the ignition timing

Service Checklist

3 Check and reset distributor points. Replace if necessary.

 4 Check and adjust ignition timing.

5 Adjust valve clearances to specification.

Incorrect ignition timing costs money. It is possible to spend a great deal more than is necessary over a given period of motoring time, say 8000 km, simply because fuel is being wasted by ignition timing that is not at the optimum setting. It is also possible that a severe case of incorrect ignition timing, if allowed to continue over a long period, may cause damage to moving parts of the engine. So regular adjustment of the timing is an essential step in servicing your car.

The petrol engine is powered by heat generated in its cylinders by the combustion of a petrol/air mixture, the means of ignition of this mixture being an electric spark. The exact moment of combustion is critical, because the piston must be in the best possible position when the ignited mixture begins to force it down its cylinder.

If combustion occurs too soon, the pressure tries to force the upwards moving piston downwards before it is ready to change direction. This means that the engine is working against itself, causing the temperature inside the combustion chamber to rise unnecessarily. Subjected to this overheating for too long, the crown of the piston becomes burned, as do its valves, and the engine loses compression and power.

This condition is known as pre-ignition, and can be identified by a knocking sound from the engine and a tendency to run on after it has been switched off.

If combustion occurs too late, the piston has already begun its journey towards the bottom of the cylinder. The energy in the fuel is not properly used, fuel is wasted and overheating occurs. The engine sounds flat and lacks power.

Combustion is not instantaneous. It takes perhaps 0.003sec. So it is timed to begin a fraction of a second before the piston reaches the top of its travel, known as top dead centre or TDC. This fraction of time is converted, for practical purposes, into a number of degrees of crankshaft rotation.

Ignition timing is carried out to ensure that ignition occurs at precisely the correct fraction of a second. It is done by adjusting the position of the distributor (which controls the flow of sparks to each cylinder in turn) in relation to the distributor cam on the engine (which is automatically synchronized with the positions of the pistons).

If the distributor is turned one way in relation to its cam, ignition will be advanced—that is, will occur earlier. If it is turned the other way, ignition will be retarded.

Since all the pistons in an engine are linked to the same crankshaft, they can move in relation to each other only in a predetermined pattern. So once the timing has been set correctly for one piston, the others are also correct.

There is, however, one complication: to make the most economical use of the fuel and to gain maximum power, ignition must occur ever more in advance of TDC as the speed of the engine increases. This is because the time that the fuel takes to burn is constant but—in that same 0.003 sec—both piston and crankshaft are covering more 'ground'.

To cope with the timing alterations necessary when the engine speed is increasing, or when the throttle is only partly open, the distributor is provided with two adjusting mechanisms: the centrifugal advance, which responds to the speed of the engine, and the vacuum advance, where the strength of the vacuum depends on the engine's load.

Order of work

Before starting work on ignition timing, you need to know three things:

1. Which of your cylinders is No. 1.
2. Which cylinder is used for the timing procedure. (In nearly all cases it is No. 1, but there are rare exceptions where the rearmost cylinder (No. 4, No. 6 or No. 8) is used instead.)
3. At what point—that is, how many degrees of crankshaft rotation before TDC—the timing cylinder should fire. (This varies from car to car, and also varies for the same car between countries which have strict emission-control regulations and those which do not.)

So ring a main distributor for your car and ask him how many degrees of advance are used for timing your car's ignition, checking at the same time that it is in fact No. 1 cylinder that is used.

A table showing how the cylinders are numbered in popular cars is on pages 35 to 41.

The actual job involves four stages:

1. Establishing, on a fixed part of the engine, a mark corresponding to TDC.
2. Establishing on a rotating part of the engine, such as the fan pulley or the flywheel, a mark to represent TDC.
3. Also establishing on the rotating component another mark—the advance mark—representing the number of degrees before TDC that combustion should take place.
4. Adjusting the distributor so that No. 1 (or other

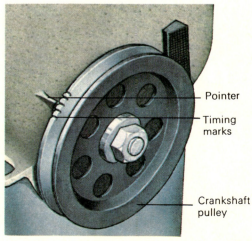

1 Ignition timing marks may be found on the crankshaft pulley at the front of the engine. Here the scale is marked on the pulley

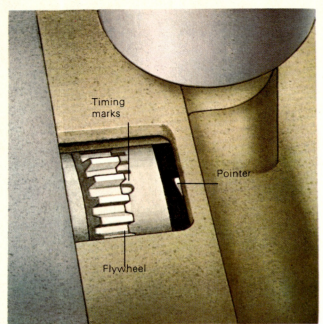

2 The timing marks may be on the flywheel, visible through a hole in the top, side or bottom of the bell housing around the clutch

appropriate) cylinder fires exactly as the advance mark on the fan pulley or flywheel lines up with the fixed mark, corresponding to TDC, which you have made on the engine.

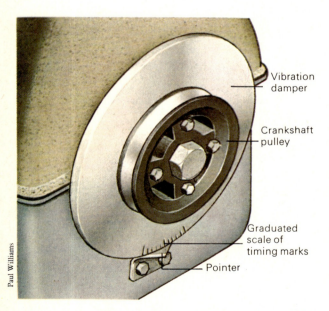

3 Some cars are fitted with vibration dampers on the crankshaft. These are mounted behind the pulley and may carry a timing scale

Finding the timing marks

The first step in timing the ignition is to find the timing marks or, if the manufacturer has failed to provide any, to make some. Figs. 1 to 3 show the most likely places for the marks to be.

If the timing marks are on the flywheel, they can usually be seen by removing a small plate bolted to the bell housing at the back of the engine. This plate is generally at the top or side, but on a few cars it is at the bottom. On some cars with transverse engines, the marks can be seen only with a small hand-mirror and a torch. Stick the mirror to the flywheel housing with a piece of chewing gum and a suitable support while you work.

Alternatively, the timing marks may be on the fan pulley or—on cars with crankshaft vibration dampers behind the fan pulley—on the dampers.

Interpreting the timing marks

The next step is to determine what the marks identify.

On the fan pulley or wherever, you may find only a single mark. If it is inscribed Top, TDC, No. 1 or 0°, it is certainly the TDC mark—the one that denotes that the cylinder used for timing purposes is in the 'fire' position. If it carries no identifying symbols, it is still almost certainly the TDC mark—except on late Volkswagens where, just to confuse the issue, it is the advance mark.

If you find two lines about 2.5 cm or 1in. apart, the one that comes up first as you rotate the engine is the advance mark, the other the TDC mark.

Some engines have two advance marks set close together. These are for use when the engine is run on different grades of fuel. The higher the octane rating (or star rating) of the fuel, the more advanced the timing will be.

On the pulley or damper, some manufacturers provide an elaborate scale, showing as many as 24° before top dead centre (BTDC) and 16° after top dead centre (ATDC) as well as the actual TDC mark. Others provide only one groove on the crankshaft fitting, but a calibrated scale bolted to a stationary part of the engine. In both cases the scale allows the timing to be varied according to the conditions under which the car may be run—altitudes higher than sea level, for example, or hot or cold climates.

Whatever marks are provided, it is a good plan to fill them carefully with white paint to make reference easier.

Making a TDC mark

Should no timing marks exist, the correct positions will have to be calculated and marked. For the TDC mark, the procedure is to get the piston in the timing cylinder (usually No. 1) exactly to the top of its compression stroke. Then you make a narrow groove in the rim of the fan pulley or damper, and make a corresponding mark precisely in line with it on an adjacent, stationary part of the engine.

To do this, the engine must be turned in its correct direction of rotation. This is most easily done with the spark-plugs removed—but first mark each plug lead by number (page 13) to avoid confusion later.

There are several ways of turning an engine slowly. One is to use the crank handle, in the unlikely—but happy—event that your car has one. Another is to turn the engine by pulling on the fanbelt, although this is difficult with big engines of 3 litres or more, and engines with electric or viscous-coupling fans should not be turned in this way.

Yet another method is to use a suitable socket and a long drive bar on the nut securing the pulley, or damper, to the crankshaft. A final alternative—although not on cars with automatic transmission—is to put the car into top gear, turn the steering on to full lock, and push the car by one of the protruding tyres.

Never turn the engine over backwards; you will get a false reading.

Since making the TDC mark requires that No. 1 cylinder be on the compression stroke (when both inlet and exhaust valves are shut at the same time), the rocker cover must be removed so you can see when this takes place. (On overhead camshaft engines, the camshaft cover must be removed.) No. 1 piston is on its compression stroke when the springs of the relevant valves are not compressed, but are their normal length. You can see whether they are

IGNITION

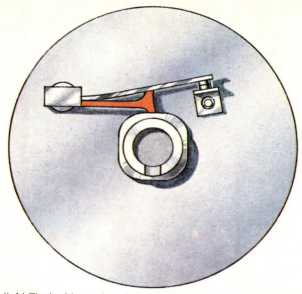

4 (left) The ignition points closed, (right) The position they must be in when the ignition is to be timed statically

simply by looking at adjacent springs, some of which will be in noticeable states of compression.

To get an exact reading of TDC, however, it is necessary to smear the threads of No. 1 cylinder's spark-plug hole with a good bubble-blowing liquid, such as detergent or a strong soap solution. To keep the bubble a manageable size, screw in the spark-plug about half to one turn or, alternatively, use a tapered wood plug with a groove cut down one side.

As you rotate the engine, the bubble will increase all the time that the piston is rising on the compression stroke. Once it has passed TDC, the bubble will shrink or burst.

Repeat the bubble-making procedure a few times to get the feel of where TDC is. You will need to turn the engine so that other pistons reach TDC before No. 1 will return to that position.

Once you are sure that you have the bubble at its maximum height, make your initial marks as described above.

Making the timing mark

Next, you want a timing mark the right number of degrees —as advised by your main agent—in advance of the TDC

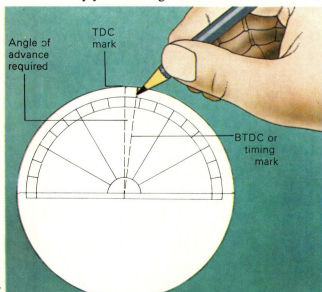

5 A timing card can be made simply by cutting a circle of card the same size as the pulley and marking on it the advance angle

mark.

There are two ways of doing this. One is to measure the diameter of the fan pulley with a ruler and cut a circle of card exactly the same size as the pulley. Then you measure out the required angle with a protractor (fig.5) and transfer the timing mark to the pulley.

For the other method, start by measuring the pulley diameter. Multiply this by 22. Then multiply by the number of degrees of ignition advance required. Divide the result by 7 x 360 (2520). This will give the required distance around the pulley rim. The formula is:

$$\text{Distance} = \frac{\text{Pulley dia.} \times 22 \times \text{advance angle}}{7 \times 360}$$

Finally, use a flexible steel tape to measure the distance around the pulley rim.

Whichever method you use, remember that as the engine rotates the timing mark must come up *ahead* of the TDC mark. As you face the engine from the fan end, for example, the direction of rotation will be clockwise, so the new mark will be to the right of TDC. The new mark is BTDC, before top dead centre.

Setting the ignition timing

There are three methods of setting the ignition timing:

1. The static method, carried out with the engine stopped. This method can be used on any car, but it is not precise, and final adjustments must be done later under road-test conditions.

2. The stroboscopic light method, carried out with the engine running at a pre-determined number of revs. It is more precise than static timing, but works only on cars with a tachometer (rev counter).

3. Monolithic timing. This relatively new method works only on cars whose crankshafts are made specially to suit it.

Adjusting the points

Before you begin checking the ignition timing, the contact-breaker points must be adjusted (pages 18 to 23).

Static ignition timing

The static method of ignition timing uses a 12-volt bulb in the middle of a length of electrical flex. This is connected across the contact-breaker points in the distributor

27

6 With the spark-plug leads numbered and removed, undo the nuts or screws securing the rocker cover and withdraw it carefully. If the gasket is bowed, leaking or damaged, it must be replaced

or coil. With the ignition on, the engine is turned over by hand so that the bulb lights up at the instant at which the relevant spark-plug would normally produce a spark.

By this means it is possible to keep adjusting the distributor until the light starts coming on when the BTDC mark on the flywheel (or wherever) is precisely aligned with the fixed reference mark. If the light comes on before the piston is at BTDC, the ignition is advanced and must be retarded; if it comes on after BTDC, the ignition is retarded and must be advanced.

Ready-made 12-volt bulb kits can be bought from most motor accessory stores. Alternatively, you can make up your own by soldering a length of electrical flex to each terminal on a 12-volt bulb, and then soldering a crocodile clip to the other end of each wire.

The procedure for static timing is in figs. 6 to 13, but two extra aspects must be borne in mind

1. If the timing marks are on the flywheel, or if the engine is to be turned by rocking the car in gear, you will probably need an assistant.

2. When you come to make the final adjustments to the timing by means of the vernier scale (fine adjuster) on

7 To find TDC, remove No 1 spark-plug and smear the hole with a good bubble-blowing liquid, in this case strong soap solution. Make sure the spark-plug hole is completely clear of all grit

8 Unless the liquid produces a very elastic bubble, it may not be necessary to insert a spark-plug to control the bubble's volume, but with thinner solutions be careful not to burst it

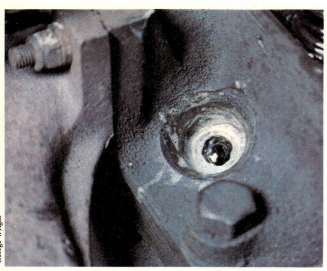

9 Once you have the feel of the process, rotate the engine until the bubble is at its greatest volume. The engine should never be turned over backwards as this will give a false reading

10 Choose a suitable stationary part of the engine near the crankshaft pulley (you may have to make a mark instead), and then mark the pulley rim exactly opposite this, using a hacksaw

IGNITION

the distributor, a maximum of six clicks in either direction is all that is allowed—even though most distributors will move 11. Any more than six and you should revert to rotating the distributor body.

Ignition timing by strobe light

A stroboscopic timing light uses a neon, or similar, high voltage bulb that triggers a flash of very short duration virtually instantly it is triggered. This flash is triggered by current in the high-tension lead to No. 1 spark-plug. So each time No. 1 sparks, the bulb lights up.

To use a strobe light, you shine it on the timing marks (on the crankshaft pulley or damper, or on the flywheel) while the engine is running. Since the strobe flashes only when No. 1 fires, it makes the pulley (or whatever) appear stationary. Hence it is easy to see whether the advance mark on the pulley is exactly in line with the fixed mark on the engine.

But there is one small problem. The number of degrees BTDC that No. 1 fires is not a constant; it varies with the speed of the engine. (Two mechanisms, the centrifugal advance and vacuum advance, control this variation.) So to use a strobe you need to know the correct dynamic ignition timing at a given speed—for example, 7° BTDC at 1000rpm—and to run the engine at the correct number of rpm. For this reason, the car must have a rev counter.

A main dealer for your car will tell you the correct advance angle and corresponding rpm for dynamic timing, and this angle should be marked on the fan pulley in the same way as if for static timing.

Stroboscopic timing lights can be hired from good motor accessory stores; for most d-i-y mechanics, buying one is an unwarranted expense.

The procedure for timing by strobe light is in figs. 14 to 18. But before you begin, it is important to disconnect the vacuum advance pipe (arrowed in fig. 11).

Safety precautions

Some strobe lights have leads which connect to the battery as well as to No. 1 spark-plug. When using these, always connect first to the battery, then to No. 1 plug. When disconnecting, always remove first the No. 1 plug leads, then the battery leads. This way you will avoid getting a high voltage shock if you touch the battery connections.

A strobe light makes everything that is rotating at

12 With the timing marks aligned and the vernier scale (if one is fitted) centred, slacken the distributor clamp bolt, located at the base of the distributor near the engine

13 The light should come on as the marks align. If it does not, rotate the distributor until it does. Now retighten the clamp on the base. Make any final fine adjustment with the vernier scale

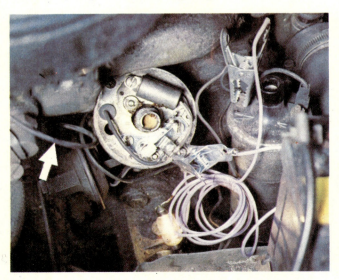

11 Remove the bakelite cap from the distributor, and connect one crocodile clip to the LT terminal on the distributor, the other to that on the coil. With the ignition on the bulb may light up

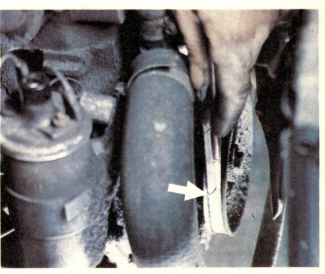

14 Stroboscopic ignition timing is done with the engine running. So it is necessary to make sure the pulley timing mark stands out. Pick it out with a thin line of white paint

15 The stationary timing mark on the front of the engine or the bell housing must be picked out with white paint. When the strobe is pointed at the two marks at speed, they must appear as one line

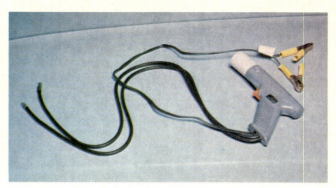

16 Some strobe lights have only two leads, but some have four. The four-lead strobes may have special triggers which allow one man to start the engine by remote control. Be careful fitting these

17 Connect the strobe to No 1 spark-plug, one lead to the plug, the other to the cap. With the vacuum pipe removed, from the carburettor, run the engine at the correct dynamic timing speed

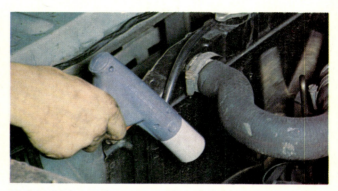

18 If the fan has not been removed, carefully shine the strobe at the timing marks. If they do not align at the correct speed, have a helper turn the distributor in either direction until they do

speed appear stationary. So keep away from moving parts. It is a wise precaution to remove the fanbelt during the period of adjustment, since the fan would be virtually invisible. Do not forget to replace it when the operation has been completed.

A further precaution, although not connected with safety, is to make sure that the BTDC mark on the pulley, damper or flywheel is clearer than all the others (for example, the TDC mark). This will prevent your setting the timing to the wrong mark. One way of doing this is in fig. 14.

If your car has timing marks on the flywheel (unless it has a transverse engine) you will need the help of an assistant.

Monolithic timing

Monolithic timing is a relatively recent development pioneered in the US by Ford. It uses a machined indicator incorporated into the crankshaft during manufacture. As the crankshaft revolves, the indicator sends an electromagnetic impulse to the monolithic timing equipment installed on the engine. This impulse triggers a timing light, and adjustment is carried out in the same manner as with a strobe light—by turning the distributor until the timing marks are aligned.

This method is claimed to be more accurate than using marks on the fan pulley or damper, since it is based on the crankshaft itself—not on a component separated from the crankshaft by a rubber ring which might distort at speed.

Electronic ignition

The two main kinds of electronic ignition—the simpler, cheaper sort which makes use of the existing contact-breaker in the ignition system, and the dearer, more sophisticated sort known as 'contactless' – both reduce petrol wastage and exhaust emission.

Where the existing contact-breaker is retained, it is possible to time the ignition statically, but most manufacturers recommend stroboscopic timing. With electronic ignition it is possible to set the ignition right up to the limit of the car manufacturer's recommended advance, because of the accuracy this type of ignition can maintain over the whole range of engine speeds. To gain the maximum benefit from it, therefore, really precise timing is needed.

Contactless ignition systems, too, can generally be timed only by strobe light. However, some systems, such as the Kenlowe Kenlomatic, incorporate their own timing light. This works in the same way as the 12-volt bulb in conventional static timing.

Road testing

Road testing is the final stage in setting the ignition timing.

Before you begin, ensure that all spark-plug leads have been replaced in their correct order and that, if you had to remove it, the fan has been replaced.

First warm up the engine to its normal working temperature. Then accelerate in top gear, and at full throttle, from about 40 km/h to 65 km/h (25mph to 40mph) while you listen for any sound of pinking—a light, metallic, knocking sound. If pinking occurs to any appreciable degree the engine is too far advanced. It should be retarded, by turning the vernier scale adjuster knob in the appropriate direction, until only the barest trace is audible.

An even better test, for which you will need an assistant, is to time the acceleration in top gear at full throttle between two fixed landmarks. Advance the ignition very slightly—that is, by one or two clicks—and test again. When the time taken to accelerate between the two points is shortest, and with only a faint hint of pinking, the timing of your car will be at its optimum.

Adjusting the valves

Service Checklist

4 Check and adjust ignition timing.

 5 Adjust valve clearances to specification.

6 Clean carburettor and adjust tick-over and mixture settings.

Adjusting the valve clearances of your car is not as complicated or as difficult as it may seem. The job requires only a few tools, some common sense and maybe up to an hour of your time.

Correct valve clearances are essential to keep your engine in a good state of tune. The only way of telling if the gaps are set correctly is to go through the procedure of checking them. Almost certainly the manufacturer of your car will have a recommended interval for this, usually between 10,000 and 15,000 km (6,000 and 10,000 miles). Also make it part of your 12,000 mile service.

The vast majority of cars on the road today use push-rod-operated ohv (overhead valve) engines. If you are not certain whether your car has this type of engine, consult the chart on pages 35 to 41.

The procedure for checking the valve clearances on an ohv engine is quite straightforward, and varies only in detail between one car and another. For methods of checking and adjustment see below. You will need a chart such as those on page 33 to adjust the clearances, or you can use the rules of five, nine or thirteen (pages 32-33).

How push-rod-operated valves work

In all four-stroke engines the camshaft is driven at half the speed of the crankshaft. It has on it a number of pear-shaped cams which indirectly open the valves in the cylinder head. Each cylinder normally has two such valves—one to permit the entry of the fuel/air mixture, the other to permit the escape of exhaust gases.

An ohv engine has small cylindrical blocks known as cam followers—the older name is tappets—located on the cams, and bearing on top of the cam followers are the push-rods. These rods, made from steel or light alloy, pass through the block up to the cylinder head where they contact one side of the rockers. The rockers are usually mounted on a tubular shaft in the cylinder head and are pivoted on this shaft in a beam fashion. Their other ends meet the tops of the valve stems.

As the cam on the camshaft reaches its peak, it pushes the cam follower upwards and the cam follower moves the push-rod in the same direction. The top of the push-rod, in turn, forces one end of the rocker arm upwards. The beam action of the rocker on its shaft means that the far end of the rocker arm then presses downwards on top of the valve stem, opening the valve inside the combustion chamber in the cylinder head.

The cam is shaped in such a way that it holds the valve open for a pre-determined period. Then, as the peak of the cam passes, the spring on the valve forces the valve shut. The rising valve stem then reverses the direction of the rocker, the push-rod and the cam follower.

Because of the extreme temperatures at which a valve operates while the engine is running, a pre-set clearance between the top of the valve stem and the rocker is essential so that the metal can expand. Although the clearance is measured in only hundredths of a millimetre or thousandths of an inch, without this gap the valve will expand to such an extent that it cannot close properly.

Similarly, it is important to make sure that the gap is correct. If the clearance is too large, the valve gear will be unduly noisy and excessive wear will occur. The engine will also develop less power, because the valve will open later and close earlier than it should.

On the other hand, a valve clearance that is too small is even more serious. In this case, the valve will not be able to close fully and eventually both the valve and the valve seat will burn out. Soon the perimeter of the valve will become so severely damaged that it will fail to make anything like a reasonable seal in the cylinder head, and serious damage to the cylinder head and perhaps the block will result.

Not all rockers are like those in fig. 5. An alternative type is that where there is no rocker-shaft as such, but where the rockers take the form of boat-like steel pressings with a central nut holding the assembly in place. The British 3-litre Ford V-six engine, for example, uses this type of rocker assembly.

Preparations

The job of adjusting valve clearances involves making a

1 On some cars you may find it necessary to remove the air-cleaner unit before you can start to check the valve clearances

2 Take a close look around the rocker box. Make sure that any pipes, wires or cables are either lifted off or disconnected

3 Rocker boxes are generally held in place by bolts or screws. This one (on a Ford Cortina) is fixed by cross-headed screws

tiny adjustment to the position of the rocker arm so that the gap between it and the valve stem is correct.

This clearance can be measured only when the valve is fully closed. But because you cannot see when this happens, you need some other method—and there are several—of telling when a particular valve is closed.

The commonest method is to use the rule of nine or rule of thirteen (see below). This is a means of telling which valve is fully closed by seeing which is fully open—if No 8 is open, for example, No 1 is closed.

But before you start work, you need to know which basic method is recommended by the manufacturer of your car. This is in the chart on pages 35 to 41.

You also need to know the correct clearances for both the inlet valves and exhaust valves—sometimes they differ—on your car. This information is usually given in a car's handbook, however skimpy.

Finally, you need to know whether the valve clearances of your car should be checked with the engine hot or cold. A 'cold' engine means that the engine has been standing idle, ie not been run, for a set period—preferably all night. Checking the clearances on a 'hot' engine requires a warming-up run of eight km (five miles). On some cars, one set of valves are checked on a hot engine and the other when the engine is cold. Alternatively, some cars can have their valve clearances checked with the engine either hot or cold, and two sets of figures are given in the handbook.

As an added complication, some manufacturers (Vauxhall are one example) recommend that the valve clearances of several of their models should be checked while the engine is ticking over. But unless the manufacturer of your car specifies otherwise, always check valve clearances with the engine stationary.

The tools you are likely to need will include either a ring spanner or a screwdriver to remove the rocker box, a set of feeler gauges, and another ring spanner and screwdriver to adjust the valve clearance. You will also need a spare rocker box gasket.

Removing the rocker box cover

To adjust the valve clearances, you may have to remove the air-cleaner unit before you can undo the rocker box retaining screws or studs. Next, a careful check around the rocker box is essential to spot any pipes, wires or other connections bolted to it. With these off and the retainers free, the rocker box can be lifted off fairly easily.

With the rocker box removed, the valve springs and the rocker arms will be exposed. Unless the valve clearance is the same for both inlet and exhaust valves, you will need to establish which are the inlet valves and which are the exhaust valves. The simplest way of doing this is to look at the two manifolds. The inlet valves will line up with the inlet manifold while the exhaust valves will line up with the exhaust manifold.

To adjust the valve clearances, you must be able to turn the engine over by hand (see page 21).

Rule of nine, rule of thirteen

On a four-cylinder in-line engine whose valves are Siamesed—that is, arranged exhaust, inlet, inlet, exhaust, and so on—the most straightforward method of valve clearance adjustment is to use the rule of nine.

To do this, you number each of the eight valves, starting from either end of the engine. Then turn the engine over until, say, No 1 valve is fully open—that is, the rocker arm has pushed the valve down as far as it will go. Next, subtract 1 from 9, which leaves 8. This means that No 8 valve is fully closed, and that is the one you adjust.

The rule of thirteen is identical in principle to the rule of nine. It applies only to in-line six-cylinder engines whose valves are arranged in the same way as that of the engine above—inlet, exhaust, exhaust, inlet and so on. As an example, if No 4 valve is fully open, subtract 4 from 13—which means you check valve No 9.

You will notice that more than one valve will be open at one time so it makes sense to check two valves (one inlet, one exhaust) together to save turning the engine over unnecessarily. When using this system, make a mark with a piece of chalk or crayon on the rocker arms because it is easy to forget which clearances have been checked.

Note, however, that the rule of nine and rule of thirteen work only on engines with Siamesed ports, and do not work on engines whose valves alternate—inlet, exhaust, inlet, exhaust, etc. Nor do they work on V or flat engines.

Adjusting 'on the rock'

Engines whose valve ports are not Siamesed are often adjusted with the valves rocking. This means you turn the engine until the valves of one cylinder begin to rock—that is, when one of the rocker arms is just beginning to move, while the other arm has just closed.

With the engine in this position, you check both the inlet and the exhaust valves of the 'opposite' cylinder in the firing order. On this chart, for a Fiat 127, the numbers refer to the *cylinders*—not to the valves:

Firing order	Valves that are rocking	Adjust
1	1	4
3	3	2
4	4	1
2	2	3

4 Having removed any retaining bolts or screws, carefully lift off the rocker box. Place it—facing upwards—on the ground

5 The difference between open and closed valves. The valve on the far left is fully open while the others are closed

6 To check the clearance of a particular valve, slide the correct feeler gauge between the valve stem and the rocker arm

VALVES

Adjusting by TDC
Yet another method, favoured by some manufacturers, is to adjust the valve clearances by bringing each piston in turn to TDC (top dead centre). First you bring No 1 to TDC on the compression stroke (page 26) and adjust both its inlet and its exhaust valves. Then you turn the engine over by 180°. This brings to TDC the next cylinder in the firing order, so now you adjust both its valves. This procedure is continued until all the valves have been adjusted.

As an example, the Talbot Avenger fires 1, 3, 4, 2, and that is the order in which you adjust the valve clearances. Note, however, that in this case too, you are adjusting by the numbers of the cylinders—not by the numbers of the valves.

Valves side by side
On engines where the inlet manifold is located on one side of the cylinder head and the exhaust manifold on the other side—in other words, cross-flow engines—the valves are sometimes arranged side by side. The best way to adjust the valve clearances is with the aid of a chart, such as this one for the Peugeot 504:

Exhaust valve fully open	Adjust inlet valve number	Adjust exhaust valve number
1	3	4
3	4	2
4	2	1
2	1	3

Should your engine be of the cross-flow type with the valves arranged in pairs, then it is possible that you will find a chart such as this in your handbook. If this is not the case, then you can use the rule of five (which is very similar in operation to the rule of nine or rule of thirteen). This will work for any four cylinder in-line cross-flow engine whose valves are arranged in this way.

Using this rule, you can adjust only *one* set of valves at a time. Start with the inlet valves. Number each inlet valve only, starting from either end of the engine. Then turn the engine until No 1 valve is fully open. Subtract 1 from 5—the result means you check inlet valve No 4. When that particular valve clearance is set, carry on and check the rest of the inlet valves in exactly the same way. Then repeat the procedure for the exhaust valves.

V engines and flat engines
With V engines and flat engines, there is no rule for working out the order to adjust the valve clearances; you need a chart showing the correct order. For example, this is the chart for the Ford V-four engine:

Valves open	Adjust
1 and 4	5 and 8
2 and 6	3 and 7
5 and 8	1 and 4
3 and 7	2 and 6

For the valve clearance order of a V-six or V-eight engine you need a table, as for a V-four. Here is a chart to cover the British Ford 3-litre V-six:

Valves that are open	Valves to check
1 and 6	10 and 7
8 and 11	5 and 4
2 and 3	9 and 12
7 and 10	6 and 1
4 and 5	11 and 8
9 and 12	2 and 3

Note the way the valves are numbered in this case:

6	12
5	11
4	10
3	9
2	8
1	7
Left hand bank	Right hand bank

Front of engine

Not all V engines number in this way. But since the valve numbering always follows the same order as the cylinder numbering, and there are two valves per cylinder, you can easily calculate one from the other. No 1 cylinder has valves 1 and 2, No 2 has valves 3 and 4, and so on.

Adjusting the valve clearances
To check the clearance of a particular valve, insert a feeler gauge of the correct thickness between the rocker arm and the valve stem. If the blade of the feeler gauge will not enter the gap, or if it is a sloppy fit, then the valve clearance will have to be altered.

With the screw and lock-nut type of fitting (fig. 8), loosen the nut and turn the adjustment screw until the feeler gauge is a reasonably tight fit between the rocker arm and the valve stem. Then hold the adjustment screw firmly in place while you tighten the lock-nut. Finally, re-check the gap.

7 To alter the valve clearance on this Ford Cortina, simply turn the adjustment nut with a good quality ring spanner

8 Many cars (like this Leyland Mini) use a screw and locknut fitting at the rocker arm. Check the clearance as in fig. 6

9 If the clearance is incorrect loosen the locknut and turn the adjustment screw. Then tighten the locknut and re-check gap

On engines whose rockers have a self-locking nut fitting (fig. 7), the procedure for checking the valve clearances is easier. Simply turn the lock-nut with a good quality ring or socket spanner until the clearance is correct. Then remove the spanner and re-check the clearance.

Once this is done, forget about the valve concerned. If you re-check later, after doing some other valve, you may find that the previously-adjusted one is now slightly out. There is nothing you can do about this.

Adjusting the valve clearances with the engine running is slightly more difficult. Long strips of metal of the correct thickness, obtainable from a main dealer or accessory store, are needed because the hammering of the rocker against the valve stem tends to flatten feeler gauge blades. Set the engine at its lowest possible idling speed by adjusting the idling screw on the carburettor. Then slide the gauge strip in and out of the gap. After a short while, you can see when the gap is momentarily at its widest, and adjustment can then be made accordingly.

Checking for wear

One possible wear point with push-rod-operated ohv valve gear is at the valve-stem end of the rocker. The end of the rocker is case-hardened but, after a considerable mileage, this hardened part of the rocker may become pierced. This means a small crater will be set up in the rocker arm and inevitably the valve stem will reach inside to the bottom of this crater. A feeler gauge slid between the two faces will tend to rest on the lip of the crater and give a false reading.

When an engine persistently rattles on one cylinder, always check for this fault. A replacement rocker is the only answer.

Valve adjustment tool

Instead of using a screwdriver and a spanner to adjust the valve clearances you could use a special all-in-one valve adjustment tool. A good example is the SPQR, which is available from most branches of Halfords and other motor accessory stores. This tool is a combined wrench and screwdriver and works by measuring the travel of the valve-adjustment screw.

If you decide to buy one, use the tool to undo the valve lock-nut, then keep turning the adjustment screw until all the clearance is taken up and the rocker arm rests on top of the valve stem. (A ratchet arrangement inside the tool will prevent you from over-tightening the screw.) Next, turn the thimble on the tool *back* the required number of 'clicks'—each click representing so many hundredths of a millimetre or thousandths of an inch. Then finally, with the tool in position, hold the thimble and tighten the lock-nut with the wrench (figs. 10 and 11).

Apart from accurately setting up the valve clearances, the manufacturer claims, the tool will take into account any wear of the valve gear. But as a final check, it is a good idea to check that the clearance is correct by briefly running the correct feeler gauge between the rocker arm and the valve stem.

The necessary socket for a SPQR adjuster is not sold with the tool itself. A *blue*-top SPQR adjuster can be used only with the special SPQR socket, but a *black*-top adjuster can be used with any 12 mm ($\frac{1}{2}$in.) drive socket.

Replacing the rocker box cover

Once all the necessary valve clearance adjustments have been made, all that remains is to make sure that the rocker box cover or covers are correctly re-fitted. Do not forget also to ensure that all the appropriate cables, wires or fittings return to their proper place.

If you have fitted a new rocker cover gasket, it is a good idea to make one further check of the retaining screws or bolts a week or so later. This will allow you to compensate for any bedding-down of the new gasket.

10 This SPQR valve adjustment tool acts as a combined wrench and screwdriver and is available from motor accessory stores

11 An SPQR adjuster can set up the valve clearances very accurately but should only be used with its accompanying chart

12 To remind yourself which of the valve clearances you have checked, it is a good idea to mark each rocker arm with a crayon

13 To change the rocker box gasket, you will need a tube of glue and a new gasket. Make sure the two gaskets match

14 Having peeled off the old gasket, clean the rocker box surface thoroughly and give it and the new gasket a thin coat of glue

15 While the two surfaces are still tacky, place the gasket into position and press it down evenly with your thumb

ENGINE DATA

Basic engine data on popular cars

The following chart supplies data on most standard models. However, because cars are occasionally built with non-standard engines, check with a main dealer if you are in any doubt about your own vehicle. Australian cars produced by Ford, Holden and Chrysler are listed separately from European models. Other European and Australian lists are combined.

MANUFACTURER	ENGINE	MODELS	FIRING ORDER	CYLINDER LAYOUT	ADJUSTING METHOD	HOT/COLD/ RUNNING	NOTES
ALFA ROMEO	Flat 4 OHC	Alfasud (all models)	1, 2, 4, 3	C	Rule of 9 (back of cam)	Cold	Check all inlets followed by all exhausts. Special tools required
	In-line 4 DOHC	Alfetta	1, 3, 4, 2	A	Rule of 9 (back of cam)	Cold	As Alfasud. Special tools required
AUDI	In-line 4 OHV	Audi 100, Coupe Mk I 1968-76 (except models with 'YM' engine)	1, 3, 4, 2	A	Adjust on rock	Warm	
	In-line 4 OHC	Audi 80 range Audi 100 Mk I ('YM'), Audi 100, Avant Mk II (except 5E models)	1, 3, 4, 2	A	Back of cam	Warm	
	In-line 5 OHC	Audi 100, Avant 5E models	1, 2, 4, 5, 3	See note	Back of cam	Warm	Cylinders numbered from radiator end of engine
BLMC	In-line 6 OHV	Austin A99 and 110 series, Austin Healey 100/6, 3000	1, 5, 3, 6, 2, 4	E	Rule of 13	Cold	
	In-line 6 OHC	Princess 2200 range, BLMC 'Six' range, Marina 2600	1, 5, 3, 6, 2, 4	E	Rule of 13 (back of cam)	Cold	
	V-8 OHV (Rover)	Rover 3500 Mk I & II, Range Rover	1, 8, 4, 3, 6, 7, 5, 2	J	See note	Hydraulic tappets—no adjustment should be necessary	
	In-line 4 OHC	Rover 2000/2200	1, 3, 4, 2	A	Back of cam	Cold	
	In-line 6 DOHC (XK Jaguar)	All 6-cylinder Jaguar models	1, 5, 3, 6, 2, 4	M	Back of cam	Cold	
	In-line 4 OHV	Land Rover (petrol)	1, 3, 4, 2	A	Rule of 9	Hot	
	In-line 4 OHV	Mini range, Moke, 1100/1300 range, 1800/1885 range, Austin A60 and Morris Oxford range, Morris Minor and Traveller, Austin Allegro 1100/1300, Wolseley 1500 and Riley 1.5, MGB/GT, Princess 1800, MG Midget and Austin Healey Sprite (to Oct. '74), Austin A35 and A40, Marina 1.3 and 1.8, Apache 1300	1, 3, 4, 2	A	Rule of 9	Cold	

Cylinder layout diagrams A, B, C, D, E, F, M

35

MANUFACTURER	ENGINE	MODELS	FIRING ORDER	CYLINDER LAYOUT	ADJUSTING METHOD	HOT/COLD/ RUNNING	NOTES
BLMC	In-line 4 OHV (Triumph)	Spitfire range, Midget (Oct. '74 to date), 1300/1500 range, Toledo, Dolomite 1300/1500, Herald range, TR2, TR3/A, TR4/A	1, 3, 4, 2	A	Rule of 9	Cold	
	In-line 4 OHC	Austin Maxi range, Austin 1500/1750, Apache 1500	1, 3, 4, 2	A	Rule of 9 (back of cam)	Cold	
	In-line 4 OHC	Triumph Dolomite 1850/Sprint, TR7	1, 3, 4, 2	A	Back of cam	Cold	
	In-line 6 OHV	Triumph Vitesse range, GT6, 2000 and 2500 range	1, 5, 3, 6, 2, 4	E	Rule of 13	Cold	
BMW	In-line 6 OHC	All 6 cylinder models	1, 5, 3, 6, 2, 4	E	Back of cam	Hot	
	In-line 4	All 4 cylinder models	1, 3, 4, 2	A	Back of cam	Hot	
CHRYSLER (UK TALBOT)	In-line 4 OHV	Hunter range, Rapier/Alpine coupe, All Minxes (incl. Super), Humber Sceptre range, 'old' Rapier/Alpine to 1967; Singer Vogue and Gazelle	1, 3, 4, 2	A	Rule of 9	Hot or Cold	Both settings usually given
	In-line 4 OHV	Avenger range	1, 3, 4, 2	A	By piston on TDC	Hot or Cold	Both settings usually given
	In-line 4 OHC	Hillman Imp range	1, 3, 4, 2	B	Back of cam	Hot or Cold	
	In-line 4 OHV	Simca 1000 range, Simca 1100 range, Simca 1301/1501, Chrysler Alpine, Chrysler 180/2 litre	1, 3, 4, 2	B	See note	Hot or Cold	Valves in balance Adjust 4 1 3 2 2 3 1 4
CITROEN	Flat 2 OHV	2CV6, Dyane 6, Ami 8, Saloon, Estate Van	1, 2	N/A	See notes	Cold	Adjust one valve of cylinder when corresponding valve on opposite cylinder is at maximum opening
	Flat 4 OHV	Ami Super, AU, GS models	1, 4, 3, 2	D	See notes	Cold	Bring the back of the cam opposite the heel of the rocker to be adjusted
	In-line 4 OHV	CX model (including diesel)	1, 3, 4, 2	B	See notes	Cold	Valve open Adjust 1ex 3in & 4ex 3ex 4in & 2ex 4ex 2in & 1ex 2ex 1in & 3ex
COLT (Mitsubishi Motor Corporation)	In-line 4	All models	1, 3, 4, 2	A	By piston at TDC	Hot	

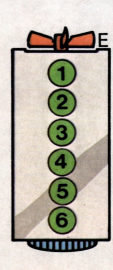

ENGINE DATA

MANUFACTURER	ENGINE	MODELS	FIRING ORDER	CYLINDER LAYOUT	ADJUSTING METHOD	HOT/COLD/ RUNNING	NOTES
DAF	Flat Twin	Daf 32, 33, 44, 46	1, 2	N/A	See notes	Cold	Each cylinder in turn with piston at TDC on compression stroke
	In-line 4 OHV	Daf 55 66	1, 3, 4, 2	B	By piston on TDC	Cold	
DATSUN	In-line 4 OHV	Sunny, Cherry, Violet 140J A/10 1-ton pick-up	1, 3, 4, 2	A	Rule of 9	Hot	
	In-line 4 OHC	C130 Laurel saloon and coupe, 510 1300 saloon, Violet J710 saloon, P510 1600 saloon, WP510 1600 station wagon all Bluebird models, all Violet 160J models	1, 3, 4, 2	A	Rule of 9	Hot	
	In-line 6 OHV	all 260 models, VP 130 Cedric 2000 station wagon	1, 5, 3, 6, 2, 4	E	Rule of 13	Hot	
	In-line 6 OHC	all other Laurel and Cedric models, 240 models, 280 models	1, 5, 3, 6, 2, 4	E	Rule of 13	Hot	
FIAT	In-line 2 OHV	Fiat 126	1, 2	N/A	Adjust on rock	Cold	
	In-line 4 OHV	Fiat 127 900cc	1, 3, 4, 2	A	Adjust on rock	Cold	
	In-line 4 OHC	Fiat 127 1050cc Fiat 128 1100 models Fiat 128 1300 3P Fiat X19	1, 3, 4, 2	A	Back of cam	Cold	
	In-line 4 OHV	Fiat 131 (all models)	1, 3, 4, 2	A	Adjust on rock	Cold	
	In-line 4 DOHC	Fiat 132, 124, 124ST, 125	1, 3, 4, 2	A	Back of cam	Cold	
	V-6 DOHC	Fiat 130	1, 4, 2, 5, 3, 6	H	Back of cam	Cold	
FORD	In-line 4 OHV	Cortina Mk I & II (except Lotus), Cortina Mk III 1300 and 1600 OHV, Cortina Mk IV 1300, Capri Mk I & II 1300 and 1600 OHV, Escort Mk I & II 1100, 1300 and 1600 OHV, Escort Mexico 1970 to 1974, Fiesta range, Zephyr 4 Mk III 100E OHV models, Anglia 1959-1968 Corsair 1500	1, 2, 4, 3	A	Rule of 9	Hot	

G: 3 6 / 2 5 / 1 4

H: 1 6 / 2 5 / 3 4

I: 4 1 / 5 2 / 6 3

J: 1 2 / 3 4 / 5 6 / 7 8

K: 2 1 / 4 3 / 6 5 / 8 7

L: 5 1 / 6 2 / 7 3 / 8 4

MANUFACTURER	ENGINE	MODELS	FIRING ORDER	CYLINDER LAYOUT	ADJUSTING METHOD	HOT/COLD/ RUNNING	NOTES		
FORD	In-line 4 OHC	Cortina Mk III 1600 and 2000 OHC models, Cortina Mk IV 1600 and 2000 OHC, Capri 1600 and 2000 OHC, Granada Mk II 2 Litre	1, 3, 4, 2	A	Back of cam	Hot			
	In-line 4 DOHC	Escort RS series	1, 3, 4, 2	A	Back of cam	Hot	Valves open 1 & 4 2 & 6 5 & 8 3 & 7	Adjust 8ex & 5in 3ex & 7in 1ex & 4in 6ex & 2in	
	V-4 OHV	Consul 2000 1972-77, Transit V4 (petrol), Corsair V4, Zephyr V4 1966 to 1972	1, 3, 4, 2	F	See note	Hot			
	In-line 6	Zephyr 6 and Zodiac Mk III to 1966	1, 5, 3, 6, 2, 4	E	Rule of 13	Hot			
	V-6 OHV	Consul 2500/3000 1972-77, Granada 3000 1972-77, Zephyr 6 and Zodiac 1966-72	1, 4, 2, 5, 3, 6	I	See note	Hot	Valves open 1 & 6 8 & 11 2 & 3 7 & 10 4 & 5 9 & 12	Adjust 10ex & 7in 5ex & 4in 12ex & 9in 1ex & 6in 8ex & 11in 3ex & 2in	
	V-6 OHV	Granada Mk II 2.3 and 2.8	1, 4, 2, 5, 3, 6	I	Adjust on rock	Cold			
HONDA	Flat-twin OHC	Honda Z, 600, TW 360 Pick-up/Van	N/A	N/A	By piston at TDC	Cold			
	In-line 4 OHC	Civic 1200/1500, Accord 1600	1, 3, 4, 2	A	Back of cam (rule of 9)	Cold			
LADA	In-line 4 OHC	All models	1, 3, 4, 2	A	Back of cam	Hot			
LANCIA	Flat 4 OHV	Flavia 2000	1, 3, 2, 4	C	By piston on TDC	Cold			
	V-4 DOHC	Fulvia (all models)	1, 3, 4, 2	C	Back of cam	Cold			
	In-line 4 DOHC	Beta (all models)	1, 3, 4, 2	A	Back of cam	Cold	Special tools required		
	Flat 4 OHC	Gamma (both models)	1, 3, 2, 4	C	Back of cam	Cold			
LOTUS	In-line 4 DOHC	Elan 1962-73 Plus 2 1967-74 Europa 1971-75 Cortina Lotus 1963-69 Escort twin cam 1969-71	1, 3, 4, 2	A	Back of cam	Cold			
	In-line 4 DOHC (slant four)	Jensen-Healey 1972 to date Lotus Elite 1974 to date Lotus Eclat 1975 to date Lotus Esprit 1976 to date	1, 3, 4, 2	A	Back of cam	Cold			

ENGINE DATA

MANUFACTURER	ENGINE	MODELS	FIRING ORDER	CYLINDER LAYOUT	ADJUSTING METHOD	HOT/COLD/ RUNNING	NOTES
MAZDA	In-line 4 OHC	1000, 1300, B1600, B1800, 1000 Hatchback, 1300 Hatchback 818, 616, 929	1, 3, 4, 2	A	Rule of 9	Hot	
MERCEDES	Information not available						
MORGAN	In-line 4 OHC	4/4 1600	——As for Ford 1600 OHV engine——				
	V-8 OHV	Plus 8	——As for Rover 3500 OHV engine——				
MOSKVICH	In-line 4 OHC	All OHC petrol models	1, 3, 4, 2	A	Back of cam	Hot	
OPEL	In-line 4 OHC (with rockers)	Manta range 1600, 1900 1970-77 (except cars with hydraulic tappets) Ascona range 1600 and 1900 1975-77	1, 3, 4, 2	A	Rule of 9	Cold	
PEUGEOT	In line 4 OHC	104 range, 304 range	1, 3, 4, 2	B	Adjust on rock	Cold	
	In-line 4 OHV	504 range (incl. diesel models)	1, 3, 4, 2	B	See note	Cold	Valve open Adjust 1ex 3in, 4ex 3ex 4in, 2ex 4ex 2in, 1ex 2ex 1in, 3ex
	V-6 SOHC (per bank)	604	1, 6, 3, 5, 2, 4	G	See note	Cold	As Renault 30TS
POLSKI-FIAT	In-line 4 OHV	Polski Fiat all models	1, 3, 4, 2	A	Rule of 9	Hot	
PORSCHE	Flat 4 OHV	914 series, 912 E, 912 range, 356 range	1, 4, 3, 2	F	By piston on TDC	Cold	
	Flat 6 OHC	911 series, 914/6 2.0	1, 6, 2, 4, 3, 5	I	By piston on TDC	Cold	
	In-line 4 OHC	924 2.0	1, 3, 4, 2	A	Back of cam	Hot	
RELIANT	In-line 4 OHV	Kitten, Robin, Regal range (1962 onwards)	1, 3, 4, 2	A	Rule of 9	Hot or Cold	Both settings should be given
	V-6 OHV	Scimitar	1, 4, 2, 5, 3, 6	I	See note	Hot	Adjusting method the same as for Ford Granada 3000 Mk I
RENAULT	V-6 SOHC (per bank)	30 TS	1, 6, 3, 5, 2, 4	G	See note	Cold	With No. 1 piston on TDC adjust inlet valves 1, 2, 4 and exhaust valves 1, 3, 6. Turn crankshaft one full revolution and adjust 3, 5, 6 inlet valves and 2, 4, 5 exhaust valves

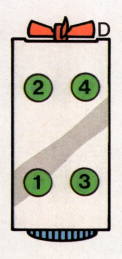
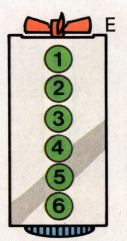

MANUFACTURER	ENGINE	MODELS	FIRING ORDER	CYLINDER LAYOUT	ADJUSTING METHOD	HOT/COLD/ RUNNING	NOTES
RENAULT	In-line 4 SOHC	R.14 models	1, 3, 4, 2	B	Rule of 9	Hot or Cold	
	In-line 4 OHV	Renault 4, 5, 6, 8, 10, 12, 15, 16 and 17 models	1, 3, 4, 2	B	Rule of 9	Hot or Cold	
	In-line 4 OHV	Renault 20 model	1, 3, 4, 2	B	Rule of 9	Cold	
SAAB	V-4 OHV	Saab 95/96 V4	1, 3, 4, 2	F	See note	Cold	With No. 1 piston on TDC, adjust valves, 1, 2, 4, 6. With No. 4 piston on TDC, adjust valves 3, 5, 7, 8
	In-line 4 OHC	Saab 99 range	1, 3, 4, 2	A	Back of cam	Cold	
SKODA	In-line 4 OHV	All models to date	1, 3, 4, 2	A	Adjust on rock	Cold	
TOYOTA	In-line 4 OHV	KE series, Corolla 'E' 30 and Liftback Hi Ace, Hi Lux, Toyo Ace, Corona KP series, Toyota 1000, Crown, Dyna (petrol), Coaster, Stout TETA series, Carina, Celica, Treuno	1, 3, 4, 2	A	Adjust on rock	Hot	
	In-line 4 OHV (Diesel)	Dyna	1, 3, 4, 2	A	Adjust on rock	Hot	
	In-line 4 OHV (Diesel)	Toyo Ace	1, 2, 4, 3	A		Running	Each valve in turn
	In-line 4 SOHC	Corona, Celica, Hi Lux, Hi Ace, Cressida (RT, RA, RX, RN, RH)	1, 3, 4, 2	A	See note	Hot	With No. 1 on TDC compression stroke adjust valves 1, 2, 3, 5. Turn crankshaft 360°. Adjust valves 4, 6, 7, 8
	In-line 4 DOHC	Celica, Treuno GT models, Celica (RA series)	1, 3, 4, 2	A	Back of cam	Cold	
	In-line 6 OHV	Land cruiser, Toyota truck	1, 5, 3, 6, 2, 4	E	Rule of 13	Hot	
	In-line 6 SOHC	Corona, Cressida, Crown (MX & MS series)	1, 5, 3, 6, 2, 4	E	See note	Hot	With No. 1 piston on TDC compression stroke, adjust inlet valves on 1, 2, 4 cylinders and exhaust 1, 3, 5. Rotate crankshaft 360°. Adjust inlet valves 3, 5, 6 and exhaust valves 2, 4, 6
VAUXHALL	In-line 4 OHV	Viva HA, HB Victor, Victor 101 and VX 4/90 1961-67, Firenza OHV	1, 3, 4, 2	A	Rule of 9	Hot or running	
	In-line 4 OHV	Viva HC Chevette, Cavalier 1256	1, 3, 4, 2	A	Each valve in turn	Running	
	In-line 4 OHC	Victor and VX 4/90 1967 to date, Magnum 1800, 2300 to date, Viva 1800, 2300 1972 to date, Firenza 1800, 2300 1972 to date, Viva 1600 1968-1972, Viva GT series, Firenza 2000SL 1968 to 1972	1, 3, 4, 2	A	Back of cam	Hot	Separate tappet adjusting screw in each tappet
	In-line 6 OHV	Ventora range	1, 5, 3, 6, 2, 4	E	Rule of 13	Hot	
VOLKSWAGEN	Flat 4 OHV	Type 1 (Beetle) range, Type 2 (Transporter) Type 3 Type 4	1, 4, 3, 2	D	By piston on TDC	Cold	

ENGINE DATA

MANUFACTURER	ENGINE	MODELS	FIRING ORDER	CYLINDER LAYOUT	ADJUSTING METHOD	HOT/COLD/ RUNNING	NOTES
VOLKSWAGEN	In-line 4 OHC (with rockers)	Polo, Derby, Golf, Golf L	1, 3, 4, 2	A	By piston on TDC	Warm	
	In-line 4 OHC	Golf range (except Golf and Golf L), Scirocco range, Passat range, LT (petrol) range	1, 3, 4, 2	A	Back of cam	Warm	
	In-line 4 OHC (with rockers)	VW K70 range	1, 3, 4, 2	A	By piston on TDC	Cold	
	In-line 4 OHV	LT (diesel) range	1, 3, 4, 2	A	Adjust on rock	Warm	
VOLVO	In-line 4 OHC	244, 245	1, 3, 4, 2	A	Back of cam	Hot or Cold	
	V-6 SOHC (per bank)	264, 265	1, 6, 3, 5, 2, 4	G	By piston on TDC	Cold	
	In-line 4 OHV	343, 66	1, 3, 4, 2	B	By piston on TDC	Cold	
	In-line 6 OHV	164	1, 5, 3, 6, 2, 4	E	See note	Hot or Cold	No. 1 piston on TDC (compression stroke) adjust 1, 2, 3, 6, 7, 10. No. 6 piston on TDC adjust 4, 5, 8, 9, 11, 12
	In-line 4 OHV	142, 144, 145 (1966-74), P130 (1961-70), P120 (1961-67), P1800/ES (1961-72), P220/S (1962-69), PV544 (1960-65), P210 (1960-69)	1, 3, 4, 2	A	See note	Hot or Cold	No. 1 on TDC adjust 1, 2, 3, 5. No. 4 on TDC adjust 4, 6, 7, 8
CHRYSLER AUSTRALIA	In-line 6 OHV	Valiant 6 VC, VE, FV	1, 5, 3, 6, 2, 4	E	Hot	See note	By piston on TDC (see article), but turn over 120° between No. 1 and No. 5, No. 5 and No. 3 etc.
	In-line 6 OHV	Valiant Hemi 6 VG, VH	1, 5, 3, 6, 2, 4	E	Hydraulic tappets—no adjustment		
	V-8 OHV	Valiant VE (eng. no. VE 62288 onwards)	1, 8, 4, 3, 6, 5, 2, 7	J	Hydraulic tappets—no adjustment should be necessary		
FORD AUSTRALIA	In-line 6 OHV	Falcon 6 XK, XL, XM (except 200 cu. in.), XP (except 200 cu. in.)	1, 5, 3, 6, 2, 4	E	Each valve in turn	Running	
	In-line 6 OHV	Fairlane 6 ZD, ZF, ZG Falcon 6 XY, XA, XB, XM (200), XP (200), XR, XT, XW	1, 5, 3, 6, 2, 4	E	Hydraulic tappets—no adjustment should be necessary		
	V-8 OHV	Fairlane V-8 ZF, ZG Falcon V-8 XA, XB, XW (351 cu. in.)	1, 3, 7, 2, 6, 5, 4, 8	L	Hydraulic tappets—no adjustment should be necessary		
	V-8 OHV	Falcon V-8 XW (302 cu. in.), XR, XT Fairlane V-8 ZD	1, 5, 4, 2, 6, 3, 7, 8	L	Hydraulic tappets—no adjustment should be necessary		
HOLDEN AUSTRALIA	In-line 6 OHV	EH, HD, HR, HK, HT, HG, HQ, HJ	1, 5, 3, 6, 2, 4	E	Hydraulic tappets—no adjustment		
	V-8 OHV	Holden V-8 HK, HT and HG (253 and 308 engined models), HQ-HJ (except 350 engined models)	1, 2, 7, 8, 4, 5, 6, 3	K	Hydraulic tappets—no adjustment should be necessary		
	V-8 OHV	Holden V-8 HK, HT, HG (327, 307 and 350 engined models), HQ-HJ (350 engined models only)	1, 8, 4, 3, 6, 5, 7, 2	J	Hydraulic tappets—no adjustment should be necessary		

Economical mixture setting

Incorrectly adjusted carburettors are expensive in terms of fuel consumption and possible wear and tear on mechanical parts. Too fast an idling speed will result in a 'racing' engine which will pull against the brakes at low speed and could cause premature wear. Incorrect mixture, if allowed to persist, can result in internal maladies such as burned valves and pistons.

All conventional carburettors have externally-mounted controls so that you can adjust the speed at which the engine ticks over, and, on variable jets, the mixture supplied to the cylinders throughout the whole range of engine speeds.

These individual controls are necessary because manufacturers' tolerances in engine construction, plus differing rates of wear, mean that every engine requires slightly different tickover and mixture settings.

Carburettor settings are normally adjusted every 10,000 km (6,000 miles) as part of normal servicing.

Order of work
The method of adjusting carburettors varies greatly from one make (or model) to another. So first you need to know what type of carburettor you have—fixed-jet or variable jet, single or multiple, and the manufacturer. Then you need to identify the idling screw and the mixture screw or screws.

But the actual job of adjustment is quite straightforward. All you have to do is adjust two or more screws in the correct order.

Do not smoke while you are working on the carburettor, and make sure that you keep all tools and components spotless—dirt is fatal to efficient carburettor working.

Preliminary checks
Before beginning to adjust the carburettor, some other areas of the engine must be checked to see that they are in good working order. Check the spark-plugs, contact breaker points and air filter. Ideally, all should be renewed, or at least cleaned thoroughly. This is especially important with the air filter, as a blocked filter leads to the carburettor receiving a richer mixture.

The points gap, plug gaps, ignition timing and valve clearances should also all be checked and, if necessary, adjusted.

Next look at the condition of the carburettor itself. Start with the throttle spindle. This is the shaft that runs through the carburettor and is activated by movement of the throttle. If it is loose or sloppy in its mounts, this indicates that the carburettor has worn badly. The replacement of this item is a major task and is probably not worth doing, since it is highly likely that other major items in the carburettor have worn too. A carburettor in this state will need expensive replacement parts, and you would do better to exchange it for a rebuilt unit, available from the manufacturers or from carburettor specialists.

Check the gasket between the carburettor and the inlet manifold for air leaks by lightly dusting it with french chalk or talcum powder and running the engine. If the powder is disturbed around the joint, the gasket should be replaced. If it is faulty it will lose air, and thereby weaken the mixture.

Move on now to the throttle and choke activating mechanisms. Both should operate quite smoothly and have no appreciable slackness. Where there is a mechanical linkage, this must be frequently lubricated with light oil, while a cable must be lightly greased and free from kinks and frays.

A guide to the mixture the engine is running, and therefore to the direction in which adjustments should proceed, is the colour of the spark-plugs. Remove a couple and study the electrodes. They should be greyish. If they are black and sooty, the mixture is too rich; if they are white and spotty, the mixture is too weak. If new plugs have been fitted, study those you have just removed.

You should check the condition of the carburettor float, and change it if it is obviously distorted.

Always make sure the engine is thoroughly warmed up before adjusting the carburettor settings, and that the choke is fully home.

Never peer directly over the carburettor intake with the air filter removed and the engine running. Any blow-back in the carburettor might result in the expulsion of a jet of petrol which is painful if it gets into the eyes.

Finally, two devices simplify a great deal of the job of adjusting carburettors. One is a Colortune kit (see

---

Service Checklist

5 Adjust valve clearances to specification.

 6 Clean carburettor and adjust tick-over and mixture settings.

7 Check oil level in gearbox (manual or automatic) and rear axle and top up if necessary. Drain gearbox and axle (if directed in handbook). Attend to miscellaneous lubrication and greasing.

Using a Colortune kit

A Colortune kit is a simple and accurate means of adjusting carburation. It consists of a glass-topped spark-plug through which the colour of the flame in the combustion chamber can be determined (see page 47).

Basically, a rich mixture burns with an orange flame, while a weak mixture is bright blue. The carburettor is adjusted in the normal way until the flame is brilliant blue speckled with orange.

The mixture in some recent carburettors cannot be altered so the flame will always be weak.

1 The Colortune carburettor adjustment kit in component form

CARBURETTOR

panel), which is recommended by many car manufacturers. The other is an air-flow meter for use when balancing twin or multiple carburettors. Both devices remove much of the possibility of incorrect adjustment due to human error or bad judgement.

Fast idle adjustment

There are two main types of carburettor—fixed-jet types such as Solex, Zenith, Weber, Ford, Hitachi, Nikki, Fish, Rochester, Holley and Carter, and variable-jet types such as Stromberg, SU and Hitachi. Both kinds are dealt with in figs. 2-10.

The fast idle adjustment is common to both; other adjustment varies according to type.

The fast idle, or choke setting, is designed to speed up the normal idle when the engine is cold and to prevent the engine stalling. It is operated by the choke cable. On most carburettors when this is pulled out a small cam on the cable opens the throttle to give a faster idle by preventing it from closing to the normal idle position. As soon as the engine is warm, the choke is pushed in, the fast idle cam returns to its normal position and the standard idle takes over.

To identify the fast idle adjustment screw, get an assistant to slowly pull out the choke control inside the car while you study carefully the action of the choke linkage under the bonnet.

When you have located the cam and screw, see if there is an arrow marked on the cam. If so, this must be aligned with the screw as it is the adjustment point. If there is no arrow, pull out the choke until the jet just begins to move (on variable jet carburettors) or as far as the choke control will go (on fixed-jet carburettors). The fast idle screw is usually held by a locknut and is not spring-loaded.

Start the engine and adjust the screw to give a very fast tickover between 1,200rpm and 1,500rpm. If no tachometer (rev counter) is fitted this may be difficult to judge, but the engine should be running fast enough not to stall when cold, and the ignition light should be off.

Fixed-jet carburettors

Fixed-choke carburettors are also known as variable-

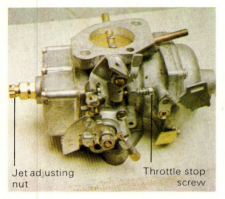
2 All Stromberg carburettors are designed with variable jets. This is the 150CD model

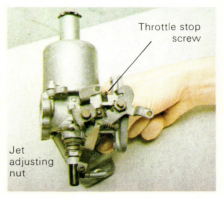
3 SU carburettors are also of the variable jet design. The HS2 model is illustrated

4 The Weber 32 DGV 15B is of the fixed-jet design and is fitted with twin venturi

5 The Zenith 361V EP carburettor is fitted to some Vauxhalls. It is a fixed-jet model

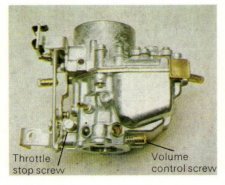
6 This Zenith 361V model differs slightly, as it is a conversion for fitment to Fords

7 The Nikki carburettor is a popular tuning modification, especially on Leyland Minis

8 A second view of the Nikki, showing the complexity of the actuating mechanisms

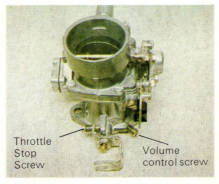
9 A simple fixed-jet design, the Solex type B30PSEI-2 is fitted to many popular cars

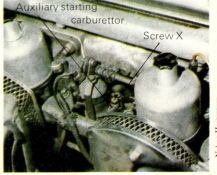
10 Twin SU HD6 carburettors fitted with an auxiliary starting carburettor

11 The fast idle choke adjustment shown in the position where the choke is fully home. The cam and adjusting screw are arrowed

12 Pull the choke out as far as it will go and you will be able to observe the action of the adjusting screw on the choke cam

13 Starting the engine from cold, pull out the choke and adjust the screw so that the idle is fast enough to prevent stalling

vacuum (or -depression) or fixed-jet (or -venturi) carburettors. All units of this type have a slow-running circuit which is arranged to supply fuel and air mixture to the engine to enable it to tick over at low speed. This circuit has an outlet below the carburettor for, whenever the engine is running slowly, a very high suction exists below the throttle and this can be used to draw the mixture from the slow-running circuit.

When the throttle is opened, the suction falls and the circuit goes out of action. The carburettor then runs on the main jet circuit, which does not require individual setting. In this way, when the idle is adjusted, the mixture supplied when the car is being driven is not altered.

To adjust a fixed-jet carburettor, you must first find its throttle stop (idling) screw. This is a spring-loaded screw which bears on, or is mounted on, the throttle lever at the carburettor base. If you have trouble in locating the throttle stop screw, the easiest method is to trace the throttle linkage or cable from the bulkhead to the carburettor. Take care to identify correctly the throttle stop screw, as there may be others bearing on the linkage which are not so close to the carburettor base. These are intermediate throttle stops, and are held by locking nuts. With the exception of the fast idle screw (see above), they should not be disturbed, as this could upset the entire adjustment of the throttle linkage.

The throttle stop screw simply controls how far the throttle can return towards the fully closed position. In this way it varies the tickover speed. Turning this screw clockwise will speed up the engine; anti-clockwise rotation will slow it down.

Next, you need to find the volume control screw. This will also be found towards the base of the carburettor, close to the fixing flange. There may be two volume screws in some cases. They are made of brass, are spring-loaded, and have knurled or serrated heads. The inner end of the screw has a tapered nose and this is used to regulate the amounts of mixture which can be drawn into the carburettor barrel. The taper is delicate, and care should be taken not to seat the screw with any force, as this may ridge the taper.

Having sorted out which screw is which, the first stage of the actual adjustment is to alter the normal idle to a brisk tickover. To provide a basic setting for this, turn the volume control screw in gently until it seats, and then unscrew it two turns.

Next, start the warmed engine and alter the throttle screw position until a fairly rapid idle results. At this stage the ignition light should be out.

The next step is to adjust the mixture. To do this, screw home the volume screw, then unscrew it two turns. Now experimentally turn the volume screw, noting the behaviour of the engine. If it slows down and becomes more unstable, reverse the direction of rotation. If it speeds up and becomes smoother, continue until a peak of smoothness is reached. You will note quite easily when the ideal point has been passed, as the engine will again begin to slow.

If there are two volume screws, each must be screwed into its seat and then screwed out by exactly the same amount, then equally adjusted until the peak of smoothness is reached.

Now that you have the mixture correct, the final stage is to return the idle to normal.

If the car has a tachometer (rev counter) and the handbook gives a recommended idling speed, alter the stop screw until the correct speed registers.

If the handbook gives no information, or the car does not have a tachometer, the idling speed can still be reset with little difficulty. The throttle stop screw should be adjusted so that the ignition light barely flickers as the engine ticks over. If the light comes on too brightly, the engine will vibrate unsteadily in its mounts and may tend to stall, particularly when you brake the car to a halt on the road. On cars with automatic transmission, the engine may stall when the gears are engaged.

If the idle is set too fast, the engine will sound as if it is racing. It may be difficult to engage first gear or, on automatics, there may be pronounced lurch when you select drive or reverse. As well as wasting fuel, too fast an idle speed can also wear out the brakes if left unchecked.

Set the screw mid-way between the two extremes. When this is done, try a final, fine adjustment of the volume control screw (fig. 9) in both directions to see if the lower speed requires a slightly different mixture setting. Adjustment is then complete.

On the volume screw there may be a point where, at ideal mixture, a small amount of rotation has no effect on the idle. If this is the case, always turn the screw to the clockwise limit of the leeway.

This basic method works with most fixed choke carburettors but one or two require some additional steps to be taken.

Fish carburettors

The American Fish carburettor (manufactured in Britain by two companies, Reece and Minnow) operates on a principle different from other carburettors. The mixture strength is altered by altering the angle of the butterfly on the throttle spindle. In this way, instead of adjusting the flow of petrol into the carburettor as in other designs, it maintains a steady flow of petrol but meters the amount of air that is allowed to enter. The main jet is also adjustable.

The procedure for Fish carburettors begins by checking the operation of the accelerator pump. This is done by removing the air filter, pumping the throttle with the engine

CARBURETTOR

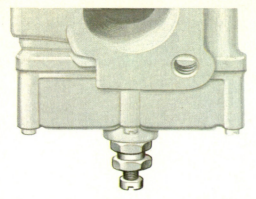

14 The jet adjusting nut on a Stromberg variable-jet carburettor

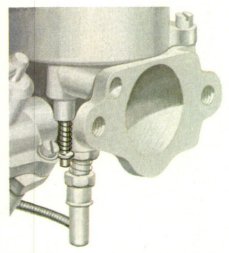

15 An SU carburettor showing the special piston-lifting pin

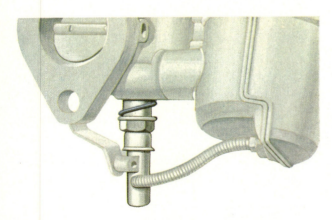

16 The jet adjusting nut and spring on an SU carburettor

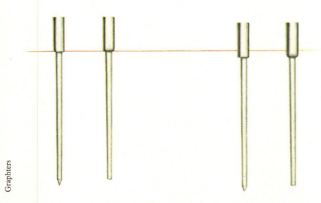

17 The correct (left), and incorrect (right), needle positions

Using a vacuum gauge

Some manufacturers, such as Ford and Weber, recommend the use of a vacuum gauge when doing carburettor adjustments.

Vacuum gauges can be used on either fixed-jet or variable-jet carburettors, and the settings are done in the usual manner. But the gauge allows you to see when the vacuum is at its greatest—the optimum point for mixture setting.

A few cars have built-in gauges on the dashboard. Others have provision for fitting a gauge in the form of a removable plug in the inlet manifold. To this you fit an adaptor (available for most cars from accessory shops), then screw the gauge on to the adaptor.

With the Weber 32 DIR 8 Mk.600 carburettor, connect the vacuum gauge and then screw the two volume screws gently home. Unscrew them by equal amounts until maximum depression (which is also the highest vacuum reading) is gained at a fast tickover. Re-adjust the tickover to the desired speed.

If a vacuum gauge is not available for this carburettor, screw home each volume screw separately, counting the number of turns necessary for each. Take an average of these and unscrew each by the average. Set a fast idle with the speed screw, then unscrew each volume screw half a turn at a time until the engine begins to run roughly. Close each by the same amount until the engine runs smoothly again. When this point is reached, reset the idle to its normal speed with the air screw.

off, and confirming that petrol squirts from the butterfly orifices. If this is not happening, the accelerator pump must be checked, which is a job for a specialist.

Next, a very fast idle of around 2000rpm must be set. Seal the air bleed hole in the end of the throttle spindle. If the engine speed remains constant, the carburettor adjustment is normal. If the engine speed increases, slacken the butterfly clamp screw, taking great care not to disturb the position of the spindle. Now turn the butterfly on the spindle to close the throttle slightly. If the engine speed falls, turn the butterfly to open the throttle slightly. Continue testing by blanking off the air bleed, after first retightening the spindle clamp screw, until there is no change in engine speed while the air bleed is covered. When this happens re-adjust the idling speed.

If you want to adjust the main jet, you will find it under a small plug in the bend in the choke tube and the float chamber. First remove the plug and, with the throttle wide open, use an Allen key to feel for the jet. Screw the jet out to enrich the mixture, in to weaken it. When the performance increase caused by this adjustment reaches its peak, set the jet just to the 'rich' side of the peak. Replace the plug and then recheck the idling mixture, as this may have been upset. Turn the screw adjuster in either direction until the most even idle is achieved.

Carburettors with air screws
Some carburettors use an air regulating screw instead of a volume control screw. These are easy to identify, as the air screw is always mounted towards the top of the carburettor, rather than the base. In such cases adjustment is carried out in the same manner as that for the usual volume screw—but whereas on the volume control system, clockwise rotation of the screw weakens the mixture, on the air

regulating system clockwise rotation enriches it. So, while the volume screw is set as far clockwise as possible, the air regulating screw is set as far anti-clockwise as possible, but not so far as to upset the idle quality.

Some other carburettors use an air screw in place of the normal throttle stop screw. These can be identified by the absence of any screw adjustment at the carburettor end of the throttle cable or linkage. The air screw meters the air flow to the carburettor, and there is a standard volume control screw to meter the fuel flow.

The Solex 26-32 DIDSA carburettor, for example, differs from other Solex carburettors in having an air screw. To adjust it, first turn the air screw until the engine is at a fast idle of about 700rpm. Then turn the volume screw, which will be the lower of the two screws, until the engine reaches a peak beyond which it will not rise regardless of the direction in which the fuel screw is rotated. Next, return the idle to 700rpm. Turn the volume screw again until the idle peaks once more and then restore the idle to 700rpm, again with the air screw. Finally, reduce the idle further until the ignition light just comes on as the engine ticks over.

Some Weber carburettors, such as the 32 DIR 8 Mk.600, also use an air screw, but are fitted with twin volume screws, one for each choke.

Twin-choke and four-choke carburettors

Some fixed-jet carburettors have two or more chokes. American carburettors, in particular, often have four. These types of carburettors are basically two or more joined together. They may be 'compound' or 'progressive' and basically are handled in the same way as single chokes.

To see if your own unit is a compound or a progressive type, ask a helper to gently depress the accelerator, with the engine off, as you watch the action of the linkage on the carburettor. If only one of the two or more throttles opens at all with the engine off, the carburettor is a compound type. If first one throttle opens, then the others, it is a progressive type. In each case, leave all but the first throttle alone, locate the idling and mixture adjustments (which will be on the first throttle only), and adjust as normally.

If both throttles open together, or all four, the unit is a true twin-choke or four-choke carburettor. In such cases, you may find only one idling screw, but there will be a volume control screw on each choke or barrel. Make the usual adjustment, but set each volume screw individually to give the smoothest tickover and highest speed. Slow the engine to the desired speed with the throttle screw and then make tiny mixture adjustments for the best idle.

Variable-jet carburettors

Variable-choke carburettors are also known as constant-vacuum (-depression) or variable-jet (or -venturi) carburettors. The choke size is varied automatically by the degree of throttle opening and the load against which the engine is working. Whether or not these carburettors have a separate idle circuit, the mixture is adjusted by the raising or lowering of the main jet.

The way that this operates is that a tapered needle (fig. 17) projects into the main jet and partially obstructs it. Raising the jet moves it to a larger section of the taper, and thus weakens the mixture; lowering the jet positions it on a smaller section of the taper, and thus causes extra richness.

The tapered needle is held in a piston or air valve. The only part of the piston assembly to come into contact with the piston chamber is the actual piston rod. All other parts of the piston assembly have sufficient clearance to prevent metal-to-metal contact. This is essential to the correct functioning of the carburettor.

The SU and Hitachi designs are similar to each other, except that where twin Hitachis are fitted, the two carburettors share one idle screw and one mixture screw. Stromberg carburettors work on the same principle as SUs and Hitachis, but in place of the internal piston, they use a rubber diaphragm and an air valve like an SU piston.

The Stromberg carburettor

On the Stromberg, the main jet is altered by an adjusting screw (fig. 14) which projects from the bottom of the carburettor, through a large hexagonal plug. To adjust the settings, begin by removing the air filter and the air valve damper. This is readily identifiable, being a large knurled plastic plug on top of the carburettor, which is removed by unscrewing. Place a pencil down the aperture from which the damper was removed, and gently depress the air valve. Next, using a small screwdriver or coin, screw up the jet adjuster nut until the jet touches the air valve. Release the pressure on the pencil and allow the air valve to return to its normal position. Now unscrew the jet three full turns.

Set the idle to around 600rpm, or until the ignition light is just flickering (the idle screw is just below the vacuum chamber which houses the air valve). Adjust the mixture screw up or down until the engine idles evenly. Test the mixture by lifting the air valve $\frac{1}{32}$in. or 0.8 mm. If the engine speed increases and stays that way, the mixture is too rich. If the engine speed falls, and stays that way, the mixture is too lean. The engine speed should pick up very slightly and then level off. If it does this, the mixture is satisfactory. Recheck the idle and top up the top of the suction chamber, known as the dashpot, with SAE 20 oil. Replace the damper and the air filter.

The SU carburettor

The procedure for setting SU and Hitachi carburettors is similar to that for Strombergs. Variable-jet Hitachis are SUs made under licence. There are some differences, however, between different SU models, in particular the various kinds of HD and HS carburettors.

On type HD SU carburettors, the main jet is totally enclosed in the base of the carburettor. For mixture adjustment it is raised or lowered by a lever which contacts the jet within the housing and is pivoted at a central point. The outer end of this lever projects horizontally from the base of the carburettor and is fitted with a spring-loaded adjusting screw which bears against a fixed lug on the carburettor. If this screw is turned clockwise, the jet will be lowered and the mixture enriched. If turned anti-clockwise, the mixture is weakened.

With this type of carburettor, take care not to confuse the mixture control with the cold start adjustment which may be mounted close by. Ensure that the adjusting screw is actually mounted on a lever which passes into the inner part of the carburettor base. Some models have automatic chokes with their own starting carburettors.

Two methods of achieving tickover are employed on the HD model. Some are fitted with ordinary throttle stop screws, while others have a slow-running volume screw.

To check if there is a throttle stop screw or a volume screw, look on the throttle linkage for any spring-loaded screw that acts on it at the carburettor end. If there is none, check for another spring-loaded screw on the carburettor, near the base. Take care not to confuse the mixture screw with this volume screw. The volume screw is usually larger than either the mixture screw or the throttle screw.

CARBURETTOR

USING A COLORTUNE KIT

18 With the engine thoroughly warm, remove No 1 spark-plug and fit the Colortune 500 plug in its place

19 Screw the plain end of the black adaptor lead on to the centre electrode of the plug, taking care not to damage the electrode

20 Slide the lower half of the viewerscope over the adaptor lead and push the clawed end over the flats of the hexagonal plug

21 Bend the adaptor lead to one side then slide it up the slot in the upper half of the scope, pushing it onto the lower half

22 With the viewerscope in position, clip the free end of the adaptor lead into the plug cap. Adjust the viewerscope mirror

23 The viewerscope is included so that any spark-plugs that are awkward to reach can be seen easily. It need not always be used

24 Start the engine and adjust the tickover to about 800 rpm. On fixed-jet carburettors this ensures the idling jet is operating

25 Start the engine and you should be able to see the flame colour. Adjust the mixture screw (or jet nut) until the flame is blue

26 The adaptor lead or viewerscope may be used to remove the hot spark-plug from the engine after loosening with a plug spanner

To adjust the HD, remove the air filter, suction chamber, piston and spring. The suction chamber is held on by two or, on larger carburettors, three set screws. Mark the chamber and its base, so it can be replaced in exactly the same position. Check that the needle is correctly mounted in the piston (fig. 17). Take care not to damage the needle, and keep the piston in a clean place.

Screw out the mixture screw until the top of the jet is flush with the jet bridge, which is visible from the top of the carburettor once the suction chamber and piston have been removed. Turn the screw clockwise by three and a half turns. Now screw in the slow running volume screw, or the throttle stop screw, to the full extent of its travel, taking care not to over-tighten. Then unscrew by two and a half turns. This provides the base setting.

Replace the piston, spring and suction chamber, and air filter. Top up the damper with SAE 20 oil to within $\frac{1}{2}$in. of the top. Start the engine and adjust the idle to around 600rpm or until the ignition light barely flickers. Check the mixture strength as detailed for the Stromberg carburettor. Some SUs have a special lifting pin on the side for this purpose (fig. 15). Adjust the mixture in the same way as on the Stromberg, bearing in mind that screwing the mixture screw clockwise will enrich the mixture, anti-clockwise will weaken it. When the mixture is correct, it may be necessary to re-adjust the idle.

Where an auxiliary starting carburettor is used, as part of an automatic choke, it is necessary to energize the starting solenoid. This is done by shorting the electrical terminal of the thermostatic switch, located on the inlet manifold, by earthing it with a screwdriver. Place the screwdriver across the terminal and one of the retaining screws to make a temporary connection. At the same time open the throttle. The carburettor will be heard to operate with a pronounced hiss. Adjust screw x (fig. 10) until the exhaust gas is just black and the idle remains unaffected. Screwing x anti-clockwise will enrich the mixture.

On SU types H and HS the procedure is broadly similar to that for the HD, but on these carburettors and on Hitachis the main jet itself protrudes from the base of the carburettor. The jet is tensioned upwards by spring loading, but it is held by a jet nut. If this nut is raised, the jet will be moved with it to a larger section of the tapered needle, and the mixture will be weakened. If the jet is lowered, it causes enrichment.

The base setting can be found in the same way as for the HD model, but turn the jet down only two turns when they are flush, or as near flush as possible, with the jet bridge. Replace the suction chamber, piston and spring as before, topping up with SAE 20 oil, then run the engine. Adjust the jet up or down until the fastest idling speed and the most even running are obtained. Again, the idle may require adjustment. The mixture strength may be tested in the same way as on the HD models.

The setting and servicing of individual carburettors is dealt with, make by make, in a series of later articles.

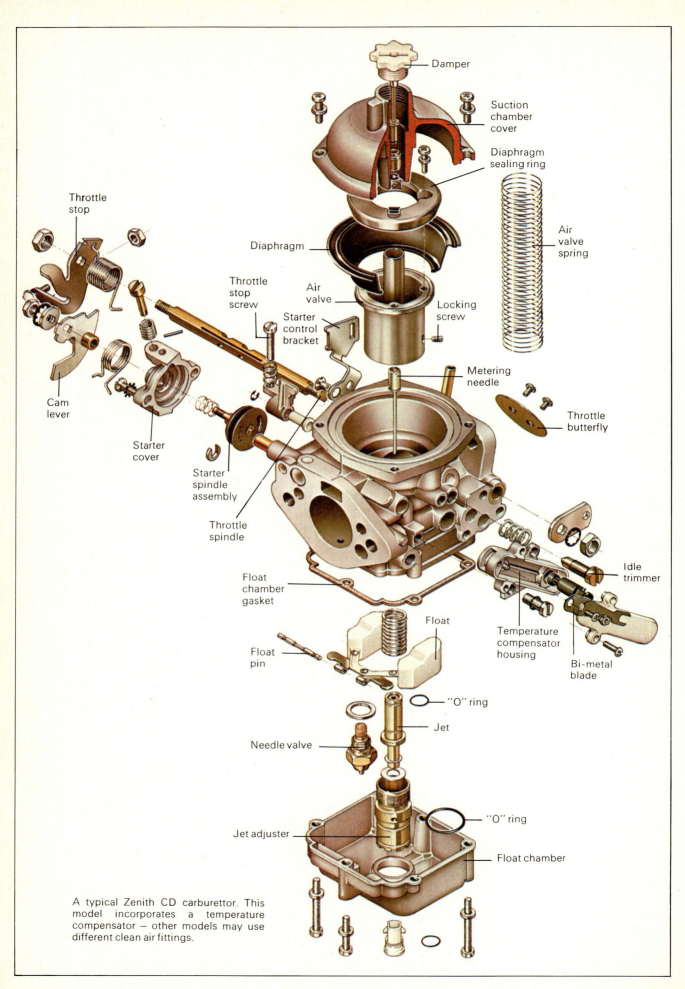

A typical Zenith CD carburettor. This model incorporates a temperature compensator — other models may use different clean air fittings.

CARBURETTOR

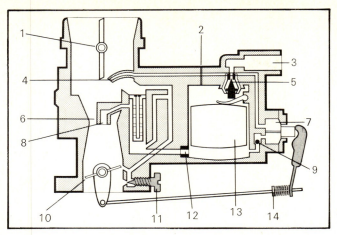

27 Diagramatic arrangement of carburettor to show component parts: 1 Open choke flap. 2 Air bleed hole. 3 Fuel supply. 4 Acceleration pump fuel nozzle. 5 Float chamber needle valve. 6 Venturi throat. 7 Acceleration pump. 8 Main fuel outlet. 9 Non-return valve. 10 Throttle butterfly. 11 Idling mixture control. 12 Main fuel jet. 13 Float. 14 Acceleration pump linkage

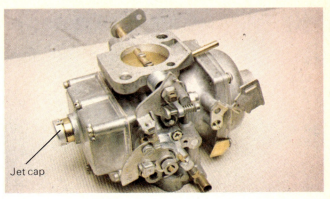

28 On Stromberg carburettors, such as this 150CD model, a cap is fitted over the jet adjustment notches to deter home adjustment

29 Early emission-control SU carburettors may be painted to indicate that they should not be altered. This is an HS4

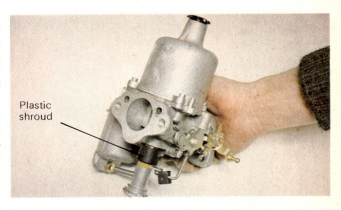

30 Later SU carburettors are fitted with plastic shrouds around the jet adjusting nut. These must be removed before adjustment

31 The plastic shroud on this SU HS6 has been removed, revealing the jet adjusting nut. The shroud cannot be replaced

Correct adjustment of the tick-over and mixture settings is even more critical where multiple, rather than single, carburettors are fitted. It is important that the carburettors are balanced and working in harmony. If the balance is incorrect they will work against one another, increasing fuel consumption and motoring costs unnecessarily.

The settings on emission-control carburettors are pre-set by the manufacturers who then usually seal off the adjustments. This setting is the optimum, but it is still usually possible to alter it if required.

For efficient operation, cars with two or more carburettors must have these properly synchronized so that each is passing the same volume of air to the engine. On most cars adjustment is possible after removing the air filters and then slackening the throttle linkage so that each unit will open independently. For a basic setting, screw out the throttle stop screws until they are just clear of their stops, indicating that both throttles are fully closed. Both screws should be turned clockwise by exactly two turns. If a brisk tickover does not result, each screw should be turned equally until it does.

The easiest way to make an airflow comparison is to insert a length of small-bore tubing just inside each carburettor intake and listen to the intake sound of the air which will be transmitted to the outer end of the tube.

Even more useful here is a stethoscope of the sort that doctors use. Comparing the sound of each carburettor will reveal whether the units are synchronized. If the sounds are not equal, the carburettor giving the louder and deeper note is passing more air than that giving a higher, quieter note. In this event, the stop screw of the louder carburettor should be slightly unscrewed and the stop screw of the quieter carburettor screwed in. In this way, by degrees, equal intake sound will be achieved.

The mixture control screws can then be individually altered for peak idle smoothness and speed. Finally, to reduce the speed to the required idle, the stop screws should be equally unscrewed. When the desired idle speed is reached, synchronization should again be checked. After slight further adjustment of the stop screws, where necessary, the linkage between the two carburettors should be carefully retightened and the air filters replaced.

If difficulty is experienced in judging the intake sounds, various air flow meters, such as the Carbalancer kit (fig. 39), are available from accessory shops. These record the flow on a scale and permit comparison. With these devices, adjustment is carried out in the normal way.

Emission-control carburettors
For some years it has been known that the chemicals contained in exhaust gas can be dangerous to health. Further research has shown that these chemicals can be kept to the lowest possible percentage by precise tuning of the carburettor to provide the weakest mixture on which the engine will run satisfactorily. The most critical phase for the production of exhaust gas pollutants is at tickover.

To supply the ideal mixture, carburettor design has been greatly adapted since the 1960s. Owners of cars built from 1970 onwards will probably find that they have an emission-control or semi-emission-control carburettor fitted.

Such carburettors still have external adjustment controls, but increasingly they are becoming blanked off since it is self-defeating to design a carburettor to provide the 'perfect' mixture and then allow this setting to be altered. The process of blanking off, usually referred to as tamperproofing, differs from manufacturer to manufacturer.

Before you consider attempting any adjustments, check whether your car is still within the warranty period. If it is, you may find that unauthorized adjustments invalidate the warranty as far as engine faults are concerned.

Legal requirements
Clean air and tamperproofing regulations are decided by Government departments and are not brought about by any action on the part of the carburettor manufacturers. Much of the following information involves the removal of the shrouds and seals designed to deter private adjustment. Currently, no penalty exists in Britain to prevent such action.

On the Continent, however, random roadside checks are carried out to assess the level of exhaust pollution when the vehicle's engine is idling. If it is found that more than 4.5 per cent of carbon monoxide gas is present in the exhaust, an on-the-spot fine may be imposed. This is not a fine for altering the pre-set adjustments, but for having a badly set mixture. Four-and-a-half per cent is a fairly high figure, however, and if the idle is set to make the mixture as lean as possible while still retaining a smooth tickover, the level of carbon monoxide should be well below the excess mark.

Special service tools for use on Stromberg and Zenith carburettors (see below) can be purchased from any Zenith main agent.

It is not possible to purchase, from any manufacturer, replacement seals or shrouds for tamperproof carburettors. To have the carburettor resealed it would be necessary to take the vehicle to an accredited carburettor specialist or to the vehicle's main dealer. However, they would simply readjust the controls using a carbon monoxide meter and then cap off the adjustment screws with colour coded caps which would indicate that adjustment had been carried out in an approved manner.

No carburettor manufacturer approves of the removal of shrouds and caps—the only manner in which the private owner can carry out adjustment. So be sure that adjustment is necessary before removing any tamperproofing. A symptom of poor running which seemingly points to the need for carburettor adjustment may be due to deterioration of some other mechanical part, such as burned or faulty valves or poor cylinder compression.

33 The Stromberg B26077 tool for jet adjustment on CD models

34 The Zenith B25243 tool for volume control screw adjustment

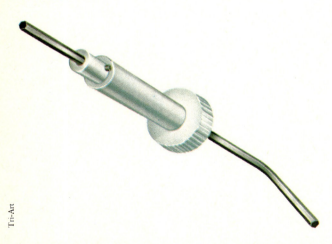

32 The B20379 adjusting tool for use on Stromberg CD carburettors

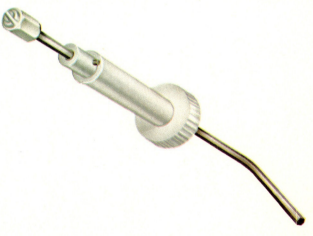

35 The Stromberg B25860 tool for the latest CD models.

CARBURETTOR

Fixed-jet emission-control carburettors

Zenith carburettors
Until the advent of tamperproofing regulations, all Zenith carburettors retained conventional adjustments. From the start of tamperproofing, Zeniths were modified by concealing the volume control screw, and in some cases the throttle stop screw, within a thin plastic shroud, covered with an aluminium cap. To gain access, the cap is punctured and service tool B25243 (fig. 34) is inserted. The outer barrel of this device unlocks a clamping ring around the actual control, while adjustment is then made with the central screwdriver device following the normal adjustment procedure as outlined on pages 42 to 44.

Ford carburettors (Motorcraft and Autolite)
Ford carburettors use no tamperproofing on the idle screw. On such units the volume screw is recessed within a built-up section of the carburettor base and is concealed beneath an aluminium cap. If this is removed, a screwdriver can be used to make normal adjustment.

Weber carburettors
Weber carburettors had no special adjustment procedure up to the advent of tamperproofing. Tamperproofed Webers have a device attached to the throttle stop screw which allows only 270° of movement. Idle speed can therefore be set only within this range of adjustment. Volume screws are recessed below a plastic cap which must be destroyed to gain access. Normal volume screw adjustment is then carried out.

French, German and Italian Solex carburettors
Far greater adaptations of design have been made on Continental Solex carburettors to produce low exhaust emissions. Although only the latest examples have been tamperproofed, clean air carburettors of unconventional idle mixture adjustment have been made for some years. The main change to look for is the fitment of a device to vary idle speed without movement of the throttle. This is achieved by an air bypass control. By altering the amount of air which can pass into the manifold via a separate channel, these operate in the manner of a throttle stop screw. When such a control is fitted, the position of the throttle should never be altered.

The air bypass screw can be located by virtue of the fact that it is a large-headed screw as opposed to the small-sized volume screw. It may be recessed into the casting, or it may project from it. It will not normally be capped off.

There will also be a volume screw near the carburettor base. In cases where two volume screws appear to be fitted, it may be assumed that only the screw that is closer to the carburettor base should be turned. The upper screw will be a control for progression mixture (just above tickover) or a fuel bleed into the bypass channel. In either case, it should not be altered in any way.

If it is clear, on a carburettor with a fixed position throttle, that the adjustment has been disturbed, a basic resetting can be made. Unscrew the throttle stop until the throttle closes completely, and then screw the stop in until a minute crack indicates the throttle is just open. It should be stressed that this is only a basic setting. It may be necessary to have the complete throttle reset by a main dealer.

36 The Carbalancer kit showing tubing, graduated scale, crocodile clamp and the two halves of the metering head

37 The two halves of the plastic metering head should be pressed together firmly. When in place, check that the inner half rotates

38 Push one end of the clear plastic tubing on to the lug on the back of the metering head, and the other end on to the flow scale

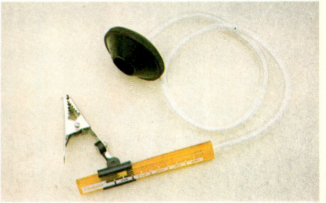

39 With the crocodile clamp clipped on to the back of the scale, the Carbalancer kit should look like this when ready for use

40 Attach the Carbalancer to a convenient point on the engine or bodywork, making sure it reaches each carburettor without strain

41 Remove the air filters. On this engine, pancake air filters are fitted but on others the arrangement may appear more complex

42 With the engine warm adjust the throttle stop screw (or with these SUs, the volume screws) equally until the idle is 1500 rpm

Japanese carburettors

Most Japanese vehicles are fitted with carburettors manufactured by Hitachi, Kiehin or Mikuni. The Japanese appear to have complied with the full spirit of the tamper-proofing regulations more than any other country. Adjustment can be made only with special tools. These are not made available to private owners, only to main dealers, so it is impossible for private owners to make any adjustments.

Variable-jet emission-control carburettors

Stromberg carburettors

On all designs of Stromberg carburettors, mixture adjustment is carried out by alteration of the main jet height or alteration of the main (metering) needle. This is mounted in the air valve. Therefore, although later Strombergs may appear to incorporate adjustments which resemble the volume screw on fixed-choke carburettors, these screws should not be disturbed.

The emphasis on Stromberg carburettors is on disguising the adjustments or by making them possible only by the use of special tools.

The CD type CDSEV carburettor, for example, is adjusted by means of a bush extending from the carburettor base into which notches are machined. CD special tool number B26077 (fig. 33) locates in these notches after a small shielding cap has been unclipped, and enables adjustment to be made as described on page 46.

Other models of CD, often known as CD3 carburettors, use an adjustable needle. Up to 1977, all carburettors so equipped could be adjusted by means of special tool number B20379 (fig. 32). This is inserted into the carburettor top after the damper has been removed. The tool has two parts. One is an outer barrel which locates in the air valve and prevents it from rotating when adjustment is carried out. A central Allen key in the tool locates in an adjuster at the bottom of the oil reservoir. When the key is turned clockwise, the needle will be raised, enriching the mixture. Anti-clockwise rotation weakens it.

There is a limit to the degree of movement on this control. Therefore, when increased resistance is felt to rotation of the key, no further attempts to turn it in that direction should be made. Do not attempt to make adjustment with only an Allen key, and not the special tool, as this may rotate the air valve and damage the diaphragm.

On very recent CD3 carburettors, the socket has been replaced with a raised blade within the oil reservoir. The mixture is adjusted with tool number B25860 (fig. 35).

Either of the special tools, B20379 or B25860, will cause the engine to run badly or stall when adjustment is made. The best procedure is to make adjustments with the

43 Disconnect the throttle linkage between the two carburettors, enabling each throttle butterfly to be independently adjusted

44 Before using the Carbalancer, the inner bezel must be rotated until it is fully open and the scale must be held vertically

CARBURETTOR

engine turned off, then to start the engine and assess the effect of the adjustment.

SU carburettors
On early SU emission-control carburettors, the alterations incorporated as an aid to exhaust cleanliness do not hamper idle adjustment. This can be carried out as described on pages 46 to 47. The mixture and throttle stop screws may have been marked with paint to indicate that they have been adjusted and sealed, but there is no reason why these settings should not be altered if the owner so wishes.

SU do not employ special tools which deter unauthorized adjustment. Their approach is to seal off the relevant controls in such a manner that any unauthorized adjustment is made obvious by damage to the shrouds. In all cases, after removal of the seals, adjustment is carried out as described on pages 46 to 47.

On tamperproof HS type carburettors, both the idle speed control (throttle stop screw) and mixture control (jet adjusting nut) are blanked off. The stop screw is recessed into a built-up part of the main body and in the top of the recess is pressed an aluminium plug. The only way, therefore, to alter tickover speed is to puncture the plug and remove it. The head of the adjusting screw will then be seen. Adjustment can be carried out in the normal manner described on pages 46 to 47.

The jet adjusting nut is housed in a two-piece plastic shroud, pressed on after initial adjustment has been made. Small pegs hold the two halves together and will have to be broken in order to remove the shroud. After this is done, adjustment can be carried out as described on pages 46 to 47.

Tamperproof HIF units have a small plastic box in which the throttle stop screw is located. This will be on the main casting, or mounted on the throttle lever, according to type. A lid on the box is clipped home after adjustment is made. To gain access to the stop screw it is necessary to break off the lid or to carefully cut off the top of the plastic box. The throttle stop screw can then be adjusted in the manner outlined on pages 46 to 47.

The mixture control on HIF carburettors is a screw, recessed into the side of the carburettor body and covered with a cap. Once again this cap must be punctured and removed to get to the mixture adjustment.

HS4C carburettors do not tamperproof the throttle stop screw. Mixture adjustment is made by movement of a pivot which acts on the main jet. The screw which moves the pivot is mounted at the base of the carburettor on the pivot bracket. It is capped off in a small box in the same way as the HIF throttle screw is tamperproofed.

46 Taking care not to disturb the adjustment made for the front carburettor, move the bezel to the orifice of the second unit

47 With the bezel in place, adjust the volume screw until there is a similar reading on the scale to that for the first carburettor

48 When the reading is satisfactory, tighten the throttle linkage carefully, then recheck the balance of the airflow on the scale

45 Insert the bezel into the orifice of the front carburettor, and, holding the outer rim, adjust the bezel to read on the scale

49 Carefully turn each volume (or throttle) screw by a similar amount to reset the idle speed and then replace the air filters

Lubrication and greasing checklist

Regular, thorough and correct lubrication is essential for preserving the life expectancy and efficiency of nearly all the mechanical components of your car.

Service Checklist

6 Clean carburettor and adjust tick-over and mixture settings.

 7 Check oil level in gearbox (manual or automatic) and rear axle and top up if necessary. Drain gearbox and axle (if directed in handbook). Attend to miscellaneous lubrication and greasing.

8 Check thickness of disc pads and brake linings. Adjust brakes where applicable.

9 Replace disc brake pads where applicable.

Inspecting oil levels (engine oil aside), topping-up procedures and general lubrication can be conveniently dealt with as part of the 12,000 mile (20,000 km) service. The gearbox and rear axle require particular attention and though these and many other components need lubricating far less regularly than was once the case there are still many parts of the car that must be attended to.

If you are unsure which parts of your car need lubricating and which lubricants to use buy a lubrication chart. These are available cheaply from oil-company franchised garages.

Gearbox and rear axle lubricants

Most gearboxes and rear axle differentials use specially developed lubricants which are designed to cope with the extremely high pressures that are created between meshing gears. These oils are sold in most motor factors and accessory shops and can be identified by the designation EP, which stands for extreme pressure. They are usually supplied in soft plastic bottles with extended nozzles which enable you to squirt the oil into the often difficult-to-reach openings in the gearbox and axle.

The grade you require may be anything from SAE 80 to 140, though EP 90 is the most common. The gearbox and rear axle may use different grades.

Some cars, for example the Chrysler Hunter, have gearboxes which run on ordinary multigrade oil, but these still demand an EP lubricant in their rear axles. On certain transverse engined front wheel drive (fwd) cars, such as the Leyland fwd models, the gearbox and final drive share a common oil supply with the engine. Multigrade engine oil is used and the gearbox is automatically topped-up and the oil is renewed at every engine oil change.

Automatic transmissions use a special type of fluid which varies from one gearbox to another. Be sure to use the correct type and grade and on no account substitute a grade other than that recommended by the manufacturer.

On some cars the transmissions are designed to be sealed for life and the oil in them will last for as long as the life span of the components. No drain or filler plugs are built into the gearbox or rear axle and they do not require any attention. The majority of transmissions, however, do have to be checked at regular intervals for correct oil level and replenished if necessary.

Topping-up the gearbox and axle

Check the specifications for the gearbox and rear axle before starting work. Wait until the engine is cold so that the oil will have run down the inside of the casing to show its true level.

It is important to get the car as level as possible so that the oil levels can be judged accurately. This may prove difficult if you have to raise the car on ramps or axle stands in order to create sufficient room to work underneath. In this case a slightly different approach is necessary. Raise the car, remove the filler plug in the way described below, overfill the component with oil and then replace the plug loosely (one or two turns of the thread is enough). Lower the car to a level position and reach underneath to remove the plug with your fingers. Let the surplus oil drain out of

1 A transverse engine, front wheel drive design (eg Mini) showing lubrication points

LUBRICATION CHECK

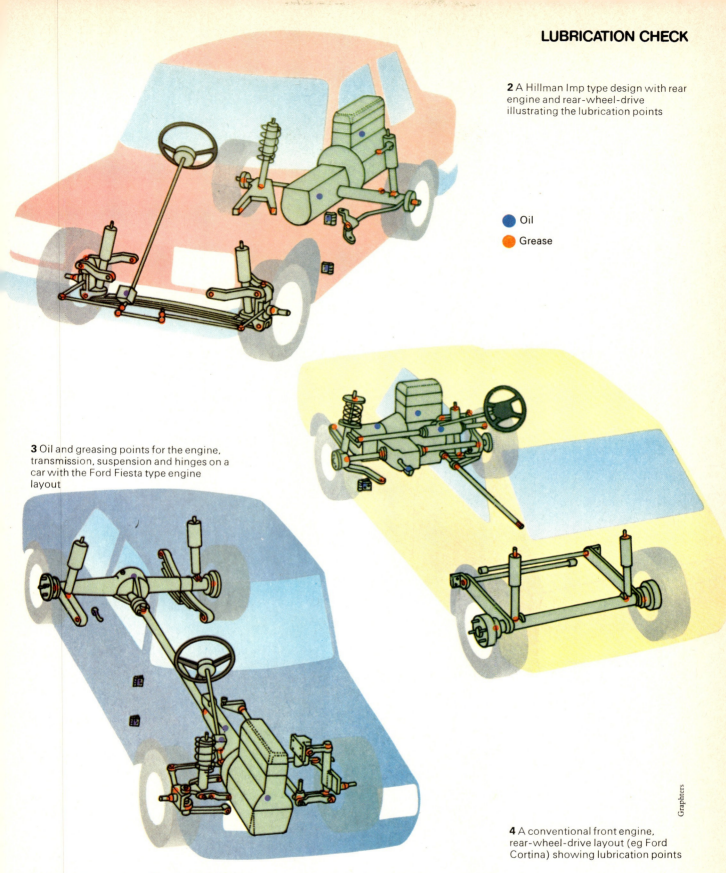

2 A Hillman Imp type design with rear engine and rear-wheel-drive illustrating the lubrication points

● Oil
● Grease

3 Oil and greasing points for the engine, transmission, suspension and hinges on a car with the Ford Fiesta type engine layout

4 A conventional front engine, rear-wheel-drive layout (eg Ford Cortina) showing lubrication points

the hole then replace the plug. Raise the car again so that you can get underneath to tighten up the plug properly with a spanner.

Most cars have a filler/level plug on the side of the gearbox (fig. 5). This is usually a hexagon-headed, bolt-shaped object that can be removed with a spanner. A similar plug may also be found on the bottom of the casing. This is the gearbox drain-plug. Do not undo this plug as it is not necessary to drain off the oil at this service interval.

Clean away all the dirt surrounding the plug before you remove it. One of the main reasons why some manufacturers fit sealed units, or advise only very infrequent servicing, is due to the damage caused by dirt and grit getting into the component through inadequate care at this point.

Remove the level plug. Fill the gearbox from the squeeze bottle until the oil level comes up to the bottom of the level-hole and just begins to run out (fig.6).

If the filler plug is on the top of the gearbox then use the dipstick to check the level. Pour the oil in slowly, a little at a time. Wait for a few moments so that the oil has a chance to run down and settle and then check the level with the dipstick. Do this until the level comes up to the right

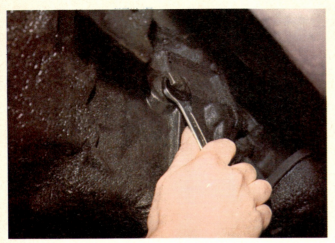

5 Find the filler/level plug on the side of the gearbox casing. It is usually bolt-shaped and hexagon-headed

6 Use the correct grade of EP oil to top up the level. Oil is usually supplied in a bottle with a nozzle to facilitate filling

mark, just as you would when topping-up the engine oil, but do not overfill.

The filler/level plug on the rear axle is easily identified. It will be found on the side of the bulbous-shaped differential housing in the centre of the axle. Top-up the axle with the correct grade of oil in exactly the same manner as described for the gearbox.

Topping-up automatic gearboxes
Some vehicles, especially those with automatic transmission, have a combined filler cap and dipstick somewhere on top of the gearbox casing. This may be located just in front of the bulkhead in the engine compartment, or through an access cover on the transmission tunnel inside the car.

Checking and topping-up automatic gearboxes involves a similar procedure but there are some extra points that must be noted. Manufacturers give different instructions for topping-up automatics depending on the type. Some specify that the job must be performed with the engine running, others specify a warm or cold engine. You may also have to select 'drive', 'park' or 'neutral' when you are noting the oil level. So check with the makers if you are unsure of the details for your car. Remember that cleanliness and the correct type of fluid are vital.

Greases and greasing
Chassis greasing and general lubrication were once a major part of a car's maintenance schedule. Every joint and moving part had to be pumped full of grease or oil at inconveniently frequent intervals. Though this is no longer the case, most cars do need periodic injections of lubricant in certain places to keep everything running smoothly.

The usual multi-purpose grease, such as Castrol LM or Shell Retinax is based on a metallic 'soap' of an element called lithium. As it has good water resistance, it is particularly useful for lubricating parts of the car that are vulnerable to poor weather conditions. It also has a high melting point and is often called HMP grease.

For some jobs, however, other types of grease may be better or even necessary. Special grease with a high molybdenum disulphide content, Duckhams MB for instance, is essential for certain universal drive joints.

Calcium-based grease, such as Castrol CL, is more runny and penetrative than its multi-purpose equivalent so it will be easier to force through joints that are blocked with old, stale grease.

Any parts of the car which need lubricating but which

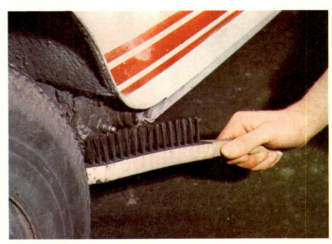

7 Clean grease nipples thoroughly to ensure that dirt or grit is not pushed into the joint during greasing

incorporate natural rubber components (in the form of bushes, boots, gaiters or similar) may require a special type of grease, such as Coparslip, if there is a likelihood of the lubricant coming into contact with the rubber. This type of lubricant should be used when greasing brake parts, which generally use rubber for seals and flexible pipes—using ordinary oily grease here could damage the rubber and have dangerous consequences.

You may find that some parts of the car appear not to have nipples fitted although they are meant to be lubricated. In these instances look for blanking plugs which cover a threaded hole. Remove these in order to screw standard grease nipples in their place. These plugs are fitted partly because they cost the manufacturer less than proper nipples and partly because they help to prevent dirt finding its way into the joints through nipples.

About four or five strokes of a grease gun (use a good quality one) will suffice to push the old lubricant out of the joint and fill it with new grease (fig. 8). You should be able to see fresh grease coming out at the edges of the joint. A rag held round the end of the gun as you apply it to the nipple can help prevent too much grease seeping into places where it is not wanted. Old, stale, solidified grease can sometimes be persuaded to move by the judicious application of heat from a blowtorch (though watch out for 'rubber' seals if you employ this method). Bear in mind that some suspension joints have to be lubricated with the weight of the car taken off them—you will need to raise the car on axle stands if this is the case.

LUBRICATION CHECK

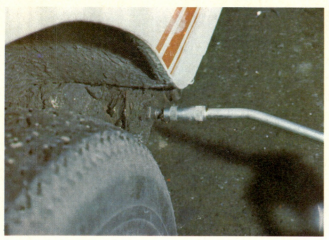

8 Apply the nozzle squarely to the nipple. Push the gun firmly to prevent it being forced away by the pressure of the grease

9 Smear the battery posts with a little petroleum jelly. This prevents the build up of chemical deposits

10 Apply a little light oil to the bonnet hinges to ensure smooth operation. Boot hinges can be treated in the same way

11 To lubricate a cable, form a funnel from grease-proof paper and tape it to the outer cable sheath about 2 cm from the end

General lubrication
To deal with most areas you need only multi-purpose grease and a can of light oil such as 3-in-1. Instead of oil you can often use a dual-purpose lubricant/water dispellent aerosol spray, WD 40 for instance.

The battery
First, though not strictly a matter for lubrication, is the battery. Lubricant is used here because at service time it is a very good idea to clean the terminals and smear them liberally with petroleum jelly (fig. 9). This does not prevent good conductivity, as would an ordinary grease, but serves very well to discourage corrosion or sulphation.

Bonnet and boot
Bonnet, boot hinges and catches need to be given a little lubrication from time to time. Multi-purpose grease is ideal where heavy applications of lubricant are needed such as on the bonnet catch, while 3-in-1 oil or engine oil will do the job for lighter duty parts such as hinges and pivots (fig. 10).

Cables (figs. 11-12)
The bonnet release is often operated by cable from a puller inside the car: this type of cable should be liberally greased. One method of doing this is to remove the inner cable and use a tin of grease such as that made by Duckhams designed for motorcycle chains—you heat the tin and immerse your cable in the molten lubricant. This impregnates the cable strands thoroughly and cuts down the need for a frequent repetition of the exercise.

Other kinds of cable, such as the clutch and some handbrake cables, can benefit from similar attention. Throttle cables almost always have an outer sheath made from a low-friction plastic material such as nylon but a little oil or aerosol spray will help the cable work smoothly, particularly if the cable is old and the inner strands are beginning to corrode. Some cars have direct mechanical throttle linkage and this can be lubricated with light oil.

There is a simple method which helps you to lubricate a cable thoroughly without the trouble of disconnecting the often inaccessible bottom end. You release the top fixing and make a little cup or funnel around the opening of the outer sheath with adhesive tape and greaseproof paper. You can then trickle as much lubricant as necessary down into the cable. Leave a little pool of oil in the funnel to soak down (figs. 11-12). You can free a stubborn sticking cable by working the inner part up and down.

Carburettor and distributor (figs. 13 to 16)
Inner components of the carburettor take care of themselves but all the external pivots and linkages should be lightly oiled. Constant depression or variable choke carburettors, for example SU and CD/Stromberg types, have damper chambers built into the tops of their piston chambers (known as 'dashpots'). These dampers need oil to function and should be topped-up regularly (figs. 13-14). SU carburettors can be lubricated satisfactorily with 20/50 or a

12 Trim off the top of the funnel and hold it upright. Pour in some light oil. This will gradually seep into the cable sheath

13 Unscrew the piston damper from the top of the carburettor taking care not to bend or damage the delicate piston head or rod

14 Top up the chamber to the bottom of the screw thread using light or multigrade engine oil. Do not overfill

15 Insert the nozzle of the aerosol spray through the holes in a distributor's baseplate to lubricate the advance weights

16 A smear of HMP grease on the centre spindle cams on which the contact-breaker heel bears, helps cut down wear

17 Lubricate door striker plates with HMP grease. Do not be too generous as a surplus will find its way on to passengers' clothes

similar engine oil, while CDs function better using a lighter oil. Zenith/Stromberg market their own product specifically for this purpose but you can use most types of light oil.

The distributor needs little lubrication but a few drops of light oil on the centre spindle (on top of the screw which is accessible after pulling off the rotor arm) and a few more on the advance weight pivots, usually accessible through the holes in the baseplate, are beneficial. These are reached by removing the distributor cap. A smear of HMP grease on the cams on which the contact-breaker fibre heel bears will help to cut down wear and prolong the interval between points adjustment.

Doors (fig. 17)
There are three distinct features of the car doors that need lubrication: the hinges, the striker plate and the lock.

Hinges can be dealt with adequately by the application of light oil or left-over engine oil to the pivots. The striker plate—the part on the door jamb with which the catch interlocks to hold the door closed—should be given a light smear of multi-purpose grease. Too much lubrication here could mean the oil will find its way on to the clothes of unwary passengers. The window winding mechanism can be dealt with too by removing the interior door trim where this is possible. This usually involves undoing a few small screws. The linkage can then be smeared with HMP grease at the pivot points.

A good lubricant for the lock mechanism is powdered graphite, which you can buy in convenient 'puffer' packs. If you cannot obtain any graphite use an aerosol spray lubricant; this has the additional advantage of preventing the lock freezing in icy weather.

Driveshafts
Driveshaft joints come in various types, some of which are packed with special greases on assembly and need not be disturbed in normal servicing. Check that the synthetic rubber boots of these joints are intact, as even a small split will let the grease out and the dirt in, quickly ruining the joint.

Other driveshaft joints such as those on the propshaft of some conventional front-engined rear wheel drive vehicles can be lubricated with HMP grease through a standard nipple. Keep oil and grease well away from the rubber of the 'spider'-type inner drive couplings such as those fitted to some fwd cars and the doughnut-shaped Rotoflex couplings found on many other cars. The gear change linkage on rear engined cars can also be lubricated with light oil at the same time while you are underneath the car.

Generators (fig. 18)
Generators need very little routine lubrication. On some dynamos (usually the Lucas ones) there is a conical housing on the rear end of the casing, which contains the rear bearing. This has a small hole in its flat rear surface. Inject the oil sparingly here because too much could contaminate the commutator and carbon brushes, and may put the device out of action. Alternators and other types of dynamo need no lubrication.

Handbrake linkages (fig. 19)
Handbrake linkages usually run underneath the car and so take some very harsh treatment from bad weather conditions. Be very generous when applying grease to pivots, cables, quadrants and any other exposed parts of the system. HMP grease can be used throughout.

Leaf springs
Leaf springs are often packed with a special graphite-laden lubricant when they are assembled and need no further lubrication. Rubber bushes in the shackles and rubber inserts between the leaves are common so keep ordinary

LUBRICATION CHECK

18 On some dynamos a small access hole is provided to lubricate the end bearing. Inject the light oil very sparingly

19 Cover handbrake cables and linkages liberally with HMP grease to protect them from bad weather conditions

20 Light oil applied to pedal pivot points ensures that the controls operate smoothly and makes the car more pleasant to drive

21 Metal seat runners and rails smeared with HMP grease make adjustment easier. Do not grease the nylon roller type

22 Steering boxes often have a filler plug on top of the casing. Clean off all dirt and lever the plug out with a screwdriver

23 Top up the steering box with the correct lubricant. This often tends to be the same oil as used for rear axles

oils and greases clear of them. If you are troubled by squeaks or creaking noises, you can squirt some old brake fluid on the springs—this is vegetable-based and designed specifically to be compatible with natural rubber, so it will not harm the rubber bushes.

Pedals (fig. 20)
Pedal pivot points are often left out of the lubrication sequence but should not be forgotten. A little oil can make a car much more pleasant to drive—there are few things so irritating as a jerky accelerator pedal or stiff clutch which makes smooth driving impossible.

Seats (fig. 21)
Seat runners and rails should be greased lightly with HMP grease. But if the runners are the plastic or nylon roller type they need no lubrication at all.

Do not use too much grease on seat parts or surplus lubricant may find its way on to the clothing of passengers.

Speedometer
If your speedometer needle tends to be twitchy and erratic, a little lubricant on the inner cable is probably all that is needed to cure it. Disconnect the outer cable from the speedometer head and pull out the inner cable. Smear the inner cable with HMP grease but leave the top 7.5 cm (3in.) clean. You risk damaging the instrument if the lubricant spreads up into it.

Steering box (figs. 22-23)
Steering boxes should be topped-up. Recommended lubricants vary but often it tends to be the same oil as used for the rear axle—this applies to all worm and roller, worm and nut and recirculating ball boxes. These types will have a filler plug located on top of the casing; it may be either square or hexagon headed.

Rack and pinion steering gear may be lubricated either with oil or grease, depending on the make and type. Check which type yours is. There may be a hexagon bolt filling point or grease nipple on top of the rack.

Ensure that the 'rubber' gaiters at each end of the rack are undamaged. Inject the correct type of lubricant into the gaiters by squirting it under the protective boots. You will probably have to release wire clips to do this.

Water pump
Many water pumps have no provision for greasing. Even on those where it is possible, you must be careful not to use too much lubricant or you may damage the carbon sealing ring dividing the bearing from the impeller. Specially formulated greases are available but ordinary HMP lithium will do the job. The lubricant is applied through a nipple or plug on the body of the pump.

Wheel bearings
Wheel bearings can be lubricated with HMP grease. On some cars you have to remove first the wheel, then the hub. This requires a very large spanner to shift the large central nut (which may be held from working loose by a split-pin). This nut may have to be retightened to a specified torque on reassembly. On newer cars you need not remove wheel or hub, but gain access to the bearing by removing the nave plate on the wheel and the dust cap beneath.

Windscreen wipers
Finally, remember to lubricate your windscreen wiper linkages. Light oil, HMP grease or glycerine can be used.

The wiper mechanism may be one of two basic types, the direct mechanical linkage or the cable driven type. Put a few drops of oil on the linkage pivots. Some cable operated wipers have grease nipples fitted to them. If so, lubricate with two or three strokes of the grease gun filled with LM grease. Do not over lubricate as the grease may get into the drive motor and damage it.

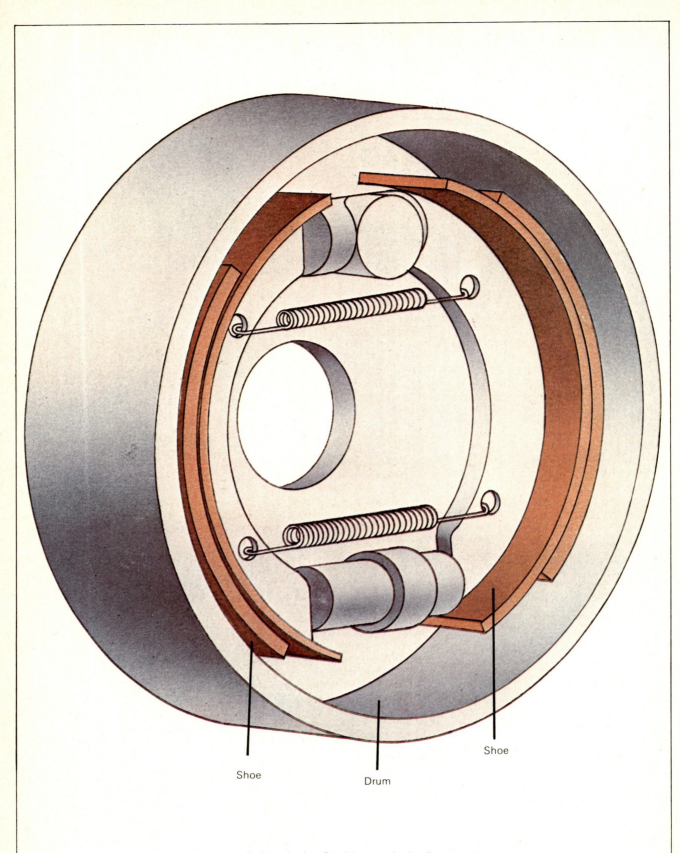

A drum brake. On this type, hydraulic pressure forces two shoes outwards to make contact with the inside face of the drum. Brakes should always be serviced in pairs to prevent the braking system from becoming unbalanced. Disc brakes are dealt with on pages 66 to 73.

Checking the brakes

From a safety viewpoint, the most important part of any vehicle is the braking system. So it is important that brakes are correctly adjusted and maintained and that, when the linings wear down, they are promptly replaced.

Replacing drum brake shoes is a 'fiddly' job, often requiring patience. But it is worth persisting with because it can save you a considerable amount of money.

The adjustment of drum brakes involves keeping the brake shoes at the correct distance from the drum. On many cars this adjustment is automatic, but on those cars which have manual adjustment, it should be checked every 1,600 km (1,000 miles).

On all types of brake, whether the adjustment is automatic or manual, the drums should be removed and the linings checked for wear every 8,000-10,000 km (5,000-6,000 miles). The linings must not be allowed to wear down to the rivets, or, where bonded shoes are fitted, below 1.5 mm ($\frac{1}{16}$ in.) from the metal shoe. The drum should also be inspected for wear, distortion or heavy scoring. If the drum is damaged it should be replaced, as it will greatly affect the effectiveness of the brakes and drastically reduce the life of the shoes.

When buying new brake shoes it is essential to get the correct replacements. So you should make a note of your car's make, model and year, or, better still, the chassis number, and quote them when buying replacement parts.

Always renew brake linings on both the back wheels, both the front wheels, or on all four. For, since new shoes will grip more efficiently than old ones, replacing shoes only on one side or corner of the car would lead to

1 The first step is to remove the wheel. When servicing brakes, do not use the car jack. Instead, use axle stands—these are safer. The wheels remaining on the ground should also be chocked

2 This Girling auto-adjust brake drum can be removed by undoing the two retaining screws. Before actually removing the drum, mark it so you can replace it the same way round on reassembly

Common types of brake adjuster

Location	Appearance	Adjuster type	Method of slackening off adjuster
Backplate	Square headed nut	Split wedge	Turn nut anticlockwise
Backplate	Lever held by a spring clip	Girling expanded pushrod spring clip	Move handbrake backwards and forwards to release tension on the spring clip. Remove clip and put handbrake off
Backplate	Protruding plastic button	Girling expanded pushrod plastic button	Locate the plastic button and pull it out. A new button should be fitted on reassembly
Behind rubber plug on backplate	Lever engaged on a toothed wheel	Lockheed cylinder	Insert two screwdrivers; with one push the lever away from the toothed wheel and with the other one turn the wheel anticlockwise
Behind rubber plug on backplate	Knurled adjuster held in position by the shoe return spring	Serrated disc	Using a screwdriver, flick the serrated nut round anticlockwise
Behind rubber plug on backplate	Two toothed plates, meshed together	Lockheed serrated plate	With a screwdriver, push adjuster plate away from the brake shoe plate
Behind rubber plug on front of drum	Slot screw	Snail cam	Insert screwdriver and turn screw anticlockwise
Behind rubber plug on front of drum	Serrated nut	Knurled disc	Same procedure as for serrated disc adjuster
Behind rubber plug on front of drum	Serrated disc supporting a lever	Girling cylinder	Insert screwdriver, lift the adjuster lever off the toothed wheel and turn the wheel anticlockwise
Next to handbrake lever on backplate	Yellow plastic clevis with black plastic pin through it	One shot	Undo the handbrake cable. Break off the black plastic pin and slide out the clevis which passes through the stop and handbrake lever

the car slewing to one side when heavily braked.

To adjust, examine and replace drum brake shoes these steps need to be followed: first the adjuster should be slackened off so that the drum can be removed and examined. Next, the shoes should be checked and replaced if necessary. Once this has been done the drum should be replaced and the brakes adjusted.

Freeing the brake drum

Before the drum can be removed it may be necessary first to slacken off the brake shoes, as these sometimes bind on the drum. This is especially true of manually adjusted brakes, but quite often automatically adjusted brakes need to have their shoes slackened off too.

If you find that the shoes are gripping the drum the adjuster will have to be located and slackened off.

As the type of adjuster fitted varies enormously among makes of car and even among models in the same range, the owner's handbook should be consulted to identify the type of adjuster fitted to your car and the procedure to be followed to slacken it off.

Failing this, the drum should be carefully inspected for signs of an adjusting nut, screw or recess in either the front or back of the drum. The table on page 61 describes the most common types of adjuster and how they can be adjusted.

3 As this drum has auto-adjusting brakes, you should be able to free it with a tap from a hammer. However, if it sticks the adjuster may have to be slackened off (see table on page 61)

4 With a finger, check the surface of the drum for scoring or any other signs of wear. Any dust in the drum should be removed with a vacuum cleaner or damp cloth so you do not inhale asbestos

6 To release the brake shoes the holding pins have to be removed. This is done by compressing and turning the dish washer with pliers while holding the pin against the backplate with a finger

7 Remove the spring and withdraw the pin from behind the back of the drum. Before going any further it is a good idea to make a note of the position of all the components to aid reassembly

9 After you have levered both shoes out of their seatings, unhook the return springs and remove the shoes completely. If the springs are worn or stretched they should be replaced

10 After placing a rubber band around the cylinder to prevent the piston from moving, clean the backplate. To avoid damaging the rubber seals, use methylated spirits or a special cleaning fluid

BRAKES

After having slackened off the adjuster the drum should be easy to remove. Remember when working on the back brakes that the handbrake should be released first.

Removing the drum

The drum can either be separate from the wheel hub or combined with it. If the brake drum has two retaining screws in between the wheel bolts, or a locating flange which holds the inside edge of the drum to the hub, then it is separate from the wheel hub. This type of drum is removed as in fig. 2.

If there is no sign of either screws or flange, then the drum is integral with the wheel hub.

To remove a brake drum which is integral with the wheel hub you need first to prise out the grease cap from the centre of the hub. You will then see a nut with castellations held in place by a split pin. After removing the pin and nut, you should then remove the spacing washer and the outer tapered roller bearing, keeping all components in their order of removal so that you know which parts to replace first when it comes to reassembly.

Some cars have a rear drum assembly which can be removed only by using a puller (fig. 12). This is a special tool which you should hire—you will not need it often enough to make it worth buying.

The puller is fitted to the wheel studs with the wheel nuts replaced to hold it in position. The centre bolt of the puller is then tightened down progressively until the drum and hub slide off the stub axle.

If you cannot hire a puller you can still remove the drum and hub assembly by replacing parts of the drum, and using the wheel to give extra leverage. You should, however, take care to exert even pressure around the wheel to prevent the distortion of any component.

You can now examine the brake linings.

Removing and replacing the brake shoes

Before removing the brake shoes note the positions of all the components so that you will be able to reassemble them correctly. If problems do occur, the drum on the other side of the car can always be dismantled and used for reference. Remember, though, that one side is a mirror image of the other, so some components will be reversed.

The brake shoes can be retained on the backplate by three different methods. Two use springs and the third uses a clip. Depending on the type of spring fitted, compress the spring and twist it or compress the spring and twist the dish washer (figs. 6 and 13). The clip type can be removed by pulling the clip downwards.

All the subsequent stages of brake shoe removal and replacement are basically the same for all makes of drum and are covered in figs. 7 to 25.

Adjusting the brakes

Automatically adjusting brakes adjust themselves in a number of different ways depending on the type fitted. Some adjust themselves after the brake pedal has been

5 Next, disconnect the handbrake linkage by removing the clevis pin, which is held in place by a split pin. The clevis pin often becomes grooved or stepped, and a worn one should be replaced

8 Next, using a shoe horn, lever off the shoes, levering from the abutment end of each shoe. A screwdriver could be used for this, but it might easily slip off and damage something

11 Check that the wheel cylinder is free to slide from side to side, by gripping it firmly and trying to move it. If it is at all reluctant to move it will need to be greased

12 Some types of brake drum require a puller to remove them. The puller is bolted on to the wheel studs, and as its central shaft is tightened down the drum is pulled off

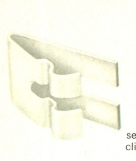

13 The spring clip is an alternative to the spring and dish washer method of securing the brake shoes. These clips can be slid off easily when compressed with pliers

depressed once, others require the handbrake to be operated several times. Some common types even require the car to be driven backwards and then stopped. In order that they are adjusted correctly the method of adjustment must be known and this information should be in the owner's manual. If in doubt consult your main dealer.

So the only attention that new ones need is to be centralized, and this is done simply by pumping the brake pedal once or twice.

The procedure for adjusting manually-adjustable drum brakes, whether fitted to front or rear wheels, is the same regardless of the type of adjuster fitted. The idea is to get

14 To grease the cylinder, lever it away from the backplate with a screwdriver and apply the grease with an old feeler gauge. Use only a high temperature grease such as Girling Blue Grease

15 After you have greased the cylinder and removed the rubber band holding the piston in place, pull out the automatic adjuster. This should then be cleaned and lightly greased

16 Before refitting the brake shoes, grease the points on the backplate where the shoes rest. Apply the grease sparingly, as none must be allowed to get on to the brake linings

17 If your old shoes look like this one, you will probably have to change the drum too. Not only has the lining been worn right away but the brass rivets have also been worn to a shine

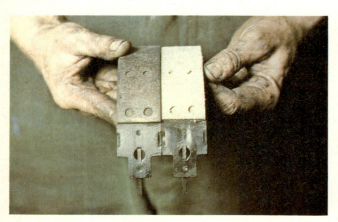

18 Before fitting new shoes, compare them with the old ones to be sure that you have the right replacements. In this case the old shoe (left) is not worn enough to need replacement

19 Whether refitting the old shoes or new replacements, give the shoe a quick rub over with a file to remove any roughness or burring of the edges, as this would impair the shoes' movement

the brakes close to a final adjustment, then to centralize them (at which point they will loosen again), and then to complete the adjustment. What you do is:

First, having determined which type of adjuster is fitted to your car, screw the adjuster in until the drum is locked—that is, until you cannot turn it.

Second, undo the adjuster slowly until you feel or hear it click twice; the drum should now be free to turn.

Third, centralize the brakes by pumping the brake pedal a couple of times; and

Fourth, repeat stages one and two.

If the adjuster does not click, note the amount the nut or screw needs to be turned for the drum to turn freely.

It is important to adjust all the brakes by the same amount so that the braking will be smooth and even.

Do not worry if the shoes bind on the drums a little; it is only because they are new. It is better to leave them slightly in contact with the drum than too loose. If they are a shade tight they will soon wear down to the correct fit. But if they are loose they could be dangerous: there will be more travel in the brake pedal before the brakes 'bite'.

With either type of brake, once the work has been completed the car should be very cautiously road tested. New shoes may give the brake pedal a spongy feel at first, as they take some time to bed in.

20 The job of replacing the shoes and springs is made much easier if you first secure one of the shoes, and then hook on the return springs. First fit the shoe attached to the handbrake

21 Next, hook the lower return spring on to the other shoe and then, using a shoe horn, lever into place the end of the shoe that seats against the wheel cylinder

22 After refitting the upper return spring, lever the upper end of the shoe into place, while keeping the springs in place. This operation is quite tricky and may take several attempts

23 As the linings of the shoes usually become dirty during the fitting operation, give them a quick rub over with a fine grade abrasive paper, such as 400 grade wet-and-dry

24 As adjusting the brakes by operating the handbrake is a long and laborious job, it can be speeded up a little by tapping the adjuster wheel round a couple of times before refitting the drum

25 After replacing the drum, refit the handbrake cable. If the two parts do not line up, do not strain the lever from the drum to meet the cable, but lengthen the adjustable end of the cable

Replacing disc brake pads

Disc brakes work like the caliper brakes on a push-bike, with friction pads instead of blocks slowing the disc and so the wheel. When the pads become worn, the cars braking will inevitably be affected and unless the pads are changed the discs themselves may be damaged.

Service Checklist

8 Check thickness of disc pads and brake linings. Adjust brakes where applicable.

★ 9 Replace disc brake pads where applicable.

10 Check and adjust handbrake.

In most cars, disc brakes are fitted only to the front wheels, because the forces under which this type of brake operates tend to make the design of the handbrake mechanism fairly complicated. Many of the more expensive and higher performance cars have disc brakes all round.

Replacing the brake pads on disc brakes is a simple job—the second time you do it. The hardest part of the job is becoming familiar with the particular brake on which you are working. And first of all you should identify which type of disc brake your car uses.

How disc brakes work

The design of disc brakes varies, but the principle on which they work is always the same. At each disc-braked wheel, a caliper unit straddles a cast-iron disc mounted on the wheel hub (or, on inboard brakes, on the axle near the differential housing). When you depress the brake pedal, a hydraulic piston or pistons make the two brake pads—one on either side of the disc—close on the disc.

Disc-brake calipers come in three main types:

Most modern cars have a fixed caliper (fig. 3), in which each pad is operated by its own piston while the caliper remains stationary.

The sliding caliper (fig. 4) has one cylinder, but two pistons. One piston acts directly on one of the pads. The other moves the yoke in the opposite direction and thus operates the second pad.

The swinging caliper (fig. 5) has only one cylinder and piston, but is pivotted in the middle and designed so that one side moves in the opposite direction to the other. Unlike the others, it uses wedge-shaped brake pads.

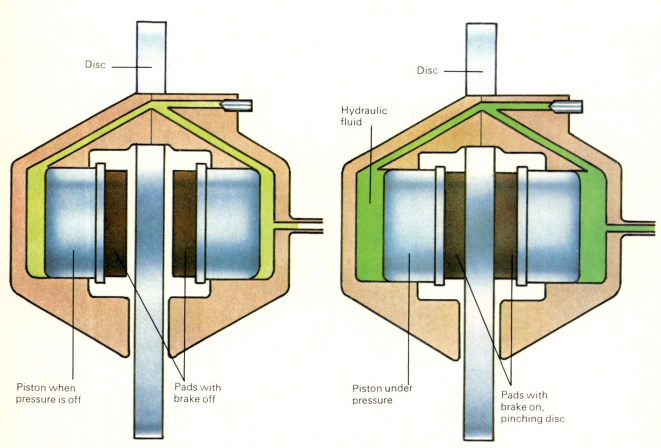

1 The basic operation of the disc brakes can be compared to those on a push bike, where the brakes' friction material sandwiches the rotating wheel. Here the brake is in the off position

2 When pressure is applied via the foot brake and the system comes under pressure, this thrusts out the disc pads against the rotating disc. The friction created will slow down the disc

DISC BRAKES

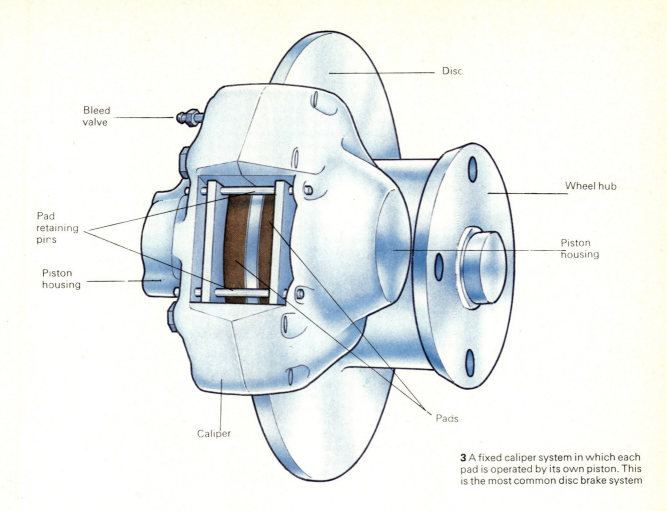

3 A fixed caliper system in which each pad is operated by its own piston. This is the most common disc brake system

Some recent dual-line braking systems, usually found on higher performance vehicles, have four-piston calipers. But the basic methods of changing the pads are the same, regardless of whether the caliper has one, two, three or four pistons.

Unlike drum brakes, disc brakes have no pull-off springs to retract the pistons after each use of the brakes. Instead, rubber seals in the caliper make the piston(s) retract. The seals also prevent the leakage of fluid past the pistons.

Disc brakes compensate automatically for brake pad wear, each piston allowing only a small running clearance between the pad(s) and the disc.

When to change pads

The condition of the pads should be checked at least every 10,000 km (6,000 miles). Inboard pads such as those fitted to the Rover 2000 wear more quickly than normal disc pads, and need to be checked more frequently.

The wedge-shaped pads found on swinging calipers should be changed when worn flat to 1.5 mm ($\frac{1}{16}$ in.).

Other pads should be changed when the lining thickness is down to 3 mm ($\frac{1}{8}$ in.).

Rear disc brakes

Some cars with rear disc brakes have handbrake pads which need to be adjusted manually—a need which is normally indicated by increased movement of the handbrake lever. Here again, regular inspection of the linings is necessary. The safe minimum thickness is 1.5 mm; wear beyond the recommended limits may damage the disc, resulting in an expensive replacement having to be carried out.

Before you begin

Before starting work, remember that in no circumstances is it necessary to separate the two halves of the caliper. So do not disturb the securing bolts at any time.

When choosing your new pads, take great care to see that they are of the correct type for your car. If you are in any doubt, quote the year, make, model and chassis number of your car when buying the replacement pads. It also helps to specify a brand name (for example, Mintex or Ferodo) as in this way you will avoid buying potentially dangerous parts.

Always buy and replace pads in sets—do not change pads singly as this might lead to dangerously uneven braking.

You will need (as well as basic tools) a jack, axle supports, wheel brace, pliers, screwdriver, wire brush and some cleaning rag. You will also need a tin of brake fluid to top up the master cylinder after changing the pads (in nearly all situations it is safe to use a universal-type fluid, except for the D series and newer Citroens which use a mineral based fluid).

The procedure for checking and replacing discs is basically the same in all cases. But with each brake system there are a few extra points, and these are detailed below.

Changing the pads

To examine the pads for wear, first remove the road wheel on which the brake is being checked. On front-wheel-drive cars, leave the car in gear when servicing the rear wheels. On rear-wheel-drive cars, leave the car in gear and also apply the handbrake when servicing the front wheels. Always chock the wheels which are not being serviced. Check both brakes on one axle at the same time so that you

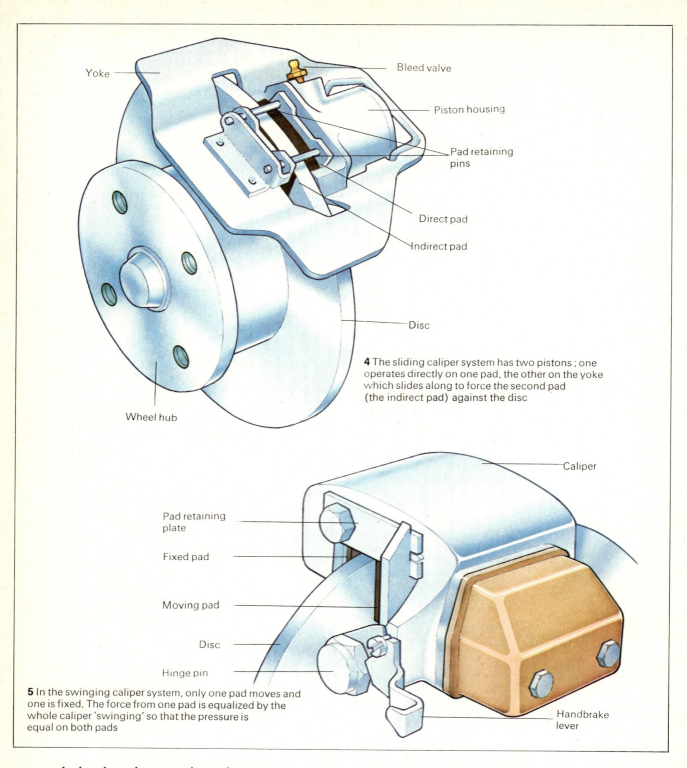

4 The sliding caliper system has two pistons ; one operates directly on one pad, the other on the yoke which slides along to force the second pad (the indirect pad) against the disc

5 In the swinging caliper system, only one pad moves and one is fixed. The force from one pad is equalized by the whole caliper 'swinging' so that the pressure is equal on both pads

can see whether the pads are wearing at the same rate.

Check through the opening in the caliper, the thickness of the friction material remaining on the pad and at the same time examine the surface of the disc itself for signs of damage, cracks or corrosion. If one side of the face of the disc is rusty then probably the piston on that side of the brake is seized. If this happens, the piston must be released and checked (see below).

To make sure that the pads are reassembled correctly, remove the pads from only one side at a time, as it lessens the risk of the pistons on the other side from coming out of their cylinders when the pistons in the caliper on which you are working are retracted.

Once the old pads are out, use a wire brush and a dry rag to clean off all the dust and dirt from the recesses in the caliper where the pads are located. Never press the brake pedal while the pads are removed, as this will force the pistons right out of their cylinders. If this happens you will most likely be faced with the long and (in this situation) unnecessary job of replacing the pistons and their seals.

Take a close look at the condition of the rubber dust covers protecting the pistons, the pistons themselves and the caliper. Check also for any fluid leaks around the pistons or hydraulic hoses. If there are any leaks, or if the other components are excessively worn or damaged, they must be replaced (see pages 80-89). Badly scored, cracked, corroded or buckled discs should also be changed.

Use an old, flat-bladed screwdriver supported on the caliper to remove all scale and rust from around the edge of the disc. Finish off the job with emery cloth.

DISC BRAKES

Releasing the fluid
When changing disc pads, the pistons in the caliper have to be retracted, and to allow the brake fluid to be displaced you have several choices. Undo the caliper bleed screw, then slowly lever back the piston which, in turn, will dribble the displaced fluid through the bleed screw. Then, when the piston is at the bottom of its cylinder, you re-tighten the bleed screw firmly.

Another method is to use a syringe or syphon to remove two-thirds of the brake fluid from the brake master cylinder, taking care not to spill any on the bodywork of the car. Then you replace the cap of the master cylinder, but arrange it so that there is enough room for any air to escape (fig. 6). In this way, when the pistons in the caliper are retracted, the displaced fluid will simply run into the master cylinder.

Of the two methods, using a syringe or syphon is the safest and simplest. It dodges the problem of a bleed screw which is rusted or hard to move and averts the possibility of air entering the brake system and of getting brake fluid on the friction surfaces of the brake pads.

To prevent the brake pads from sticking in their calipers and to facilitate the job of removing them next time, it is a good idea to file a small amount of the pad backing away. You can give the newly-filed edges a thin coating of a brake grease such as Coparslip, making sure that no grease gets on the pad linings, and smear both sides of the new shims and the back of the new pads with a squeal deterrent grease such as Disc Brake Silencer which is available from most motor accessory stores.

Once the new pads are in position and all the retaining clips and springs are reassembled, do not forget to give the brake pedal several pumps to move the pistons and pads up to the disc. While the pads are bedding in the brake pedal might feel slightly soft and spongy, but this will last only a short while.

Finally, top up the master cylinder and replace the road wheels. Tighten the nuts and remove the axle supports. Once the car is on the ground again, re-check the tightness of the wheel nuts and then give the car a brief road test.

Girling fixed caliper
The Girling fixed caliper brake is shown in fig. 3.

After cleaning the exterior of the caliper with a wire brush and a dry rag, see whether anti-squeal shims are already fitted and, if so, note their positions before removing the pad retaining clips and pins. Then gently lever the pads away from the disc with a metal lever, and pull out the shims and worn pads using pliers if necessary.

Temporarily refit one of the old pads in the caliper to retain the piston in position while the other piston is being pressed back. Lever back the other piston, ensuring that it goes slowly and evenly to prevent the piston cross-binding. Insert the new shim and pad.

Now remove the old pad from the caliper, press back that piston, fit the second new shim and pad and secure them with the retaining clips.

Girling sliding caliper (front)
The Austin Maxi and Allegro and the Peugeot 504 are among those that use the Girling sliding caliper (fig. 4).

On this disc brake the cylinder assembly is bolted rigidly to the axle and a steel yoke straddles the disc and slides in grooves in the cylinder body. Hydraulic pressure actuates both pistons, one of which acts directly on to the adjacent pad. The opposite (indirect) piston acts against the yoke, which slides in the cylinder grooves to bring the indirect pad into contact with the disc.

To change the pads, remove the anti-rattle springs, the electrical pad warning wire (if fitted), and the retaining pins. Then extract the worn pads.

If the disc is worn on one side only, then one of the pistons may be seized or the yoke may not be sliding on the cylinder. If the yoke is not sliding properly, use a hammer to gently tap the side of the yoke backwards and forwards until it moves more freely. A few drops of Coparslip and WD 40 will probably aid this.

Lever back the piston and yoke, then fit the new pads. If one of the pads has an electrical cable attached, this pad should be fitted nearer the cylinder and the cable reconnected. Repeat the procedure with the other caliper.

Girling sliding caliper (rear)
The Girling sliding yoke caliper works in basically the same way as the sliding caliper fitted to the front of various cars. But only the Peugeot 504 saloon has this design of caliper fitted to the rear brakes.

This caliper incorporates a handbrake mechanism and, because of this, certain steps must be taken when changing the pads. To fit new pads, remove the pad retaining clip and the pad wear warning lead. Unscrew the nut and bolt and remove the pad guide. Always remove the outer pad first, and then the inner one.

To disengage the handbrake mechanism, the piston acting on the inner pad must now be rotated through a set angle. There is a special tool to do this but so long as you note the exact position of the piston slot into which the pad fits then a screwdriver can be used.

Slide the screwdriver into the slot in the piston, and turn the piston until the screwdriver rests against the bottom edge of the caliper. Then, having made sure that the sliding yoke does in fact slide, push the piston back with a suitable retraction tool.

Next, push the yoke as far out from the disc as it will go and then insert the new outer pad. Use your screwdriver again to turn the piston back clockwise to its original position, and then insert the new inner pad, making sure that the pad backplate settles in the groove in the piston.

Reconnect the pad electrical wear indicator, fit the pad guide and tighten the nut and bolt. Do not forget the pad retaining spring. Repeat for the other caliper.

Girling swinging caliper
The Rover 2000, 2200 and 3500 and the Ford Zephyr and Zodiac Mk. IV models all use the Girling swinging caliper (fig. 5) which incorporates a complex handbrake arrangement.

The caliper has two friction pads, one moving and one fixed. When pressure is applied to the moving pad, the caliper reacts so that the force is equalized on both sides of the disc. The caliper is pivoted on a mounting to allow the necessary movement. The moving pad is operated by a cup assembly containing an adjustable push rod, which contacts and is moved by the cam face of the lever, which in turn is operated by either the hydraulic piston or the handbrake tappet.

To change the pads, unscrew the bolt and remove the plate and spring from the top of the brake pad holder. Swing the top of the inner pad forward and withdraw it, then pull the caliper on to the disc and remove the outer pad from the drag pins. After removing the old pads check that the handbrake lever operates smoothly and the caliper pivots easily on the pivot pin. If necessary, lubricate the bearing surfaces with Girling special grease No. 64932047.

A special tool (Girling tool No. 64932048) is now required to fit the projecting cup. (As a last resort a screwdriver can be used instead of the tool.)

The tool is operated by pushing the handle inwards to engage on the serrations and then it is turned anti-clockwise one complete turn only. On some vehicles it is necessary to disturb the brake linkage temporarily to operate the tool.

Remove the tool and clean the projecting area of the cup before smearing it with the special grease mentioned above. Refit the tool and turn the cup clockwise until it is right back and the clicking of the ratchet can be heard, or there is room to fit the new pads. Make sure the lever is vertical—if it is not, unscrew it until it is.

To prevent binding, pivot the caliper to keep a small gap between disc and tool. Before removing the tool, turn the cup until the red spot on the tool and the projecting cup tab are at 90° to each other. If the pads have been allowed to wear exceptionally thin, it may not be possible to fit the tool on the cup because of insufficient clearance between the cup and the disc. In this case, press back the cup to take up any free play and follow the procedure described—except that, in this case, the cup will have to be turned by hand until there is sufficient clearance to fit the tool.

Examine the drag pins and fit new ones if distorted. Fit the new pads in reverse order to their removal, refit the retaining plate and spring (dome uppermost), secure the tab washer. Operate the brake 20 to 30 times to settle the pads.

Lockheed swinging caliper
The Lockheed swinging caliper brake as fitted to British Leyland's 1100 and 1300, uses disc pads that are wedge-shaped when new, but which eventually wear flat.

To change the pads, depress the pad retaining spring or springs where fitted and withdraw the split pins. Rotate the pads slightly and lift them out, together with the shims if fitted (some pads lift straight out of the caliper). If the pads have two pieces of offset friction material, note their position in the caliper and also the location of any shims. If the shims are clean and undamaged there is no need to renew them.

With the piston retracted, place the new pads—with shims if fitted—into the caliper. Fit the pad-retaining spring and split pins and secure the pins by opening out their ends. Finally, slacken the pivot pin locking screw and ask a friend to press lightly on the pedal, to bring the pads into contact with the disc. Then re-tighten the bolt firmly.

Lockheed fixed caliper
The Lockheed two-piston caliper (and its variants) is fitted to several British cars, the MGB, Triumph 2000 and the Hillman Minx among others. The job of replacing the pads is much the same as with the Girling two-piston caliper but with one important exception.

The pistons both have cut-away portions which must face a certain way, but exactly which positions they adopt varies from car to car. The cut-away portion in an MGB, for example, faces towards the inner edge of the caliper, while that in a Mini Cooper faces upwards.

Over the space of 10,000 km, the pistons will rotate slightly, and if they become too far out of alignment then brake squeal and judder will certainly result.

If your car has this type of Lockheed caliper, and you do not know the correct position for the cut-away portion, your car's handbook is unlikely to give this information. Contact a reputable garage or the brake manufacturer for

6 Remove the master cylinder cap. As brake pistons are forced back into their housings, the fluid must be able to rise in the master cylinder

7 Spilt fluid will damage paint work, so protect with an old rag or remove some from the container before going any further

8 With the road wheel removed, the brake is revealed. This is a fixed caliper unit on a Ford Cortina

DISC BRAKES

9 Two retaining pins hold the pads. They may be split and bent over at the ends or, as in this case, held in place by clips

10 Remove the split or clip-held pins. If there is an anti-rattle plate over both pads, this must be removed as well

11 The brake pads and any anti-squeal shims should pull out easily now

the correct position.

Another type of heavy-duty caliper made by Lockheed was fitted to several makes of cars in the 1960s, for example the Wolseley 6/99, and is similar in principle to the DBA (Bendix) brake. With this brake the caliper must be removed from the disc to change the pads, and should then be supported to avoid straining the flexible hose.

To make sure that the caliper is mounted centrally over its disc when in position, a certain number of shims are fitted between the caliper and the mounting bracket. So you must count the shims and note their position for reference when reassembling.

Clean up the disc with a screwdriver and a piece of emery cloth, replace the pads and then bolt the caliper to the stub axle using the number of shims removed when dismantling.

ATE (Teves) fixed caliper

ATE (Teves) disc brakes are fitted mainly to German cars and in particular Mercedes-Benz and Volkswagen. The procedure for changing the disc pads is similar to that of the Girling sliding yoke caliper, but like the Lockheed two-piston brake, the piston has a cut-away portion which must be set in a certain position.

So, having removed the securing clips, the retaining clips and the cross-leaf spring, use a pair of pliers to withdraw the pads. Next, gently lever back the pistons and check the condition of the disc, the pistons and the rubber covers. When you are satisfied, use a template to check the position of the piston cut-away portion. The piston must be rotated, if necessary, so that the cut-away portion coincides with the slope of the piston template.

Finally, re-fit the pads, the cross-leaf spring, the retaining pins and the securing clips.

Akebono swinging caliper

The Akebono swinging caliper is a Japanese version of the British Lockheed swinging caliper, made under licence in Japan. This caliper is fitted to the Datsun 140J (among others). The pad thickness can be checked by removing the anti-rattle clips.

The pads, which are held in place by small recesses in their backs, locate in the caliper plate. Like their British counterparts, the pads are tapered when new.

Start by removing the anti-rattle clips and then lift out the outer pad, noting its location. Remove the shim from the pad. Next, push the caliper plate inwards (towards the chassis) and remove the inner pad—again noting its location. Remove the shim from the pad.

Press the piston to the bottom of the cylinder bore, and fit the shims to the new pads, making sure they are located correctly. Fit the inner pad into the caliper in the position from which the old one was removed. Swing the caliper outwards on the pivot and fit the outer pad into the caliper. Finally, replace the anti-rattle clip, ensuring the direction is as indicated by the label on the clip.

Sumitomo fixed caliper

The Datsun 240Z (for example) uses the Sumitomo fixed caliper which, like the Akebono swinging caliper, is a Japanese version of a British design—this time the Girling fixed body two-piston caliper. Although various models differ in size and are slightly different in appearance, the instructions for fitting new pads are similar.

First remove the hair-pin clips and slide out the pad retaining clips. Then remove the pad return spring and withdraw the worn pads and the anti-squeal shims.

12 Both these pads have been used, although only the one on the left is badly worn. To be safe, both pads must be changed

13 Use a makeshift tool to push the pistons back evenly into the caliper body to fit the new pads

14 Fit the new pads. In this case a new type has been used with a central groove for gathering harmful carbon deposits

15 Refit the retaining pins (these will centralize the pads), shims and anti-rattle plates

Press the piston to the bottom of the cylinder bore and fit the new pads and anti-squeal shims (if fitted). Fit the pad return spring and replace the pad retaining pins and hair-pin clips.

Tokico sliding caliper

Another Japanese sliding caliper that is similar to a British design is the Tokico floating caliper, fitted to the Datsun 120Y Sunny and the S2 Sunny. This caliper is similar to the Girling sliding caliper.

To change the pads, remove the wire clips from the retaining pins and then lift out the pins and the coil spring. Unhook the anti-rattle clips and withdraw the pads together with the anti-squeal shims.

Press the piston to the bottom of the cylinder bore then fit the new pads and anti-squeal shims into position. Fit the anti-rattle springs and retaining pins and secure them with the wire clip. Finally, fit the coil spring to the retaining pin furthest away from the bleed screw.

DBA (Bendix) caliper series I and II

The Renault R8 and R10 are among the cars that are fitted with the series I DBA (Bendix) caliper, although no new cars use this type. Cars that use the series II caliper include the Renault 16 (up to 1969) and the Fiat 850 sport coupe.

The calipers are of the single-piston type. The rear caliper (where fitted) incorporates an automatic handbrake adjustment mechanism.

Begin by removing the spring clip and split pin and swing the retaining blocks or clip away from the caliper, also removing the anti-rattle pads (if fitted). The caliper can now be withdrawn from the mounting bracket. Remove the pads, together with any leaf springs, but note the pad position since in some cases there is a flange at one end. Note also the position of the leaf spring.

Press the front caliper to the bottom of the cylinder bore. Where the brake is fitted to a rear wheel, a slot is provided in the centre of the caliper piston enabling the piston to be turned in a clockwise direction to retract it. When fully retracted, the slot in the piston must be positioned to accept the pad assembly. Some versions of the Fiat 850 sport coupe also have datum marks on the piston which must be aligned correctly.

Re-fit the new pads in the position from which they were removed, together with any leaf springs and replace the caliper in the mounting bracket, ensuring that the pad remains correctly located. Fit new anti-rattle pads if necessary, replace the retaining blocks or clips, and secure with spring clips or split pins.

DBA (Bendix) series III and Akebono floating caliper

The series III Bendix caliper is fitted to several different cars including the Fiat 124, 125 and 127, while the Mazda 1500 uses the Akebono caliper.

Changing the pads is fairly straightforward. Remove the fastening pins and the stoppers from above and below the caliper. Remove the caliper and anti-rattle clip. Press the piston to the bottom of the cylinder bore and remove the pads from the carrier. Fit the new pads into the carrier. Next, fit the anti-rattle clips centrally over the pads, replace the caliper over the pads, and insert the two stoppers into their respective gaps. Finally, replace the fastening pins.

Final checks

Once everything has been refitted, pump the brake pedal a few times to bring the pads up to the disc, and check the brake fluid level.

DISC BRAKES

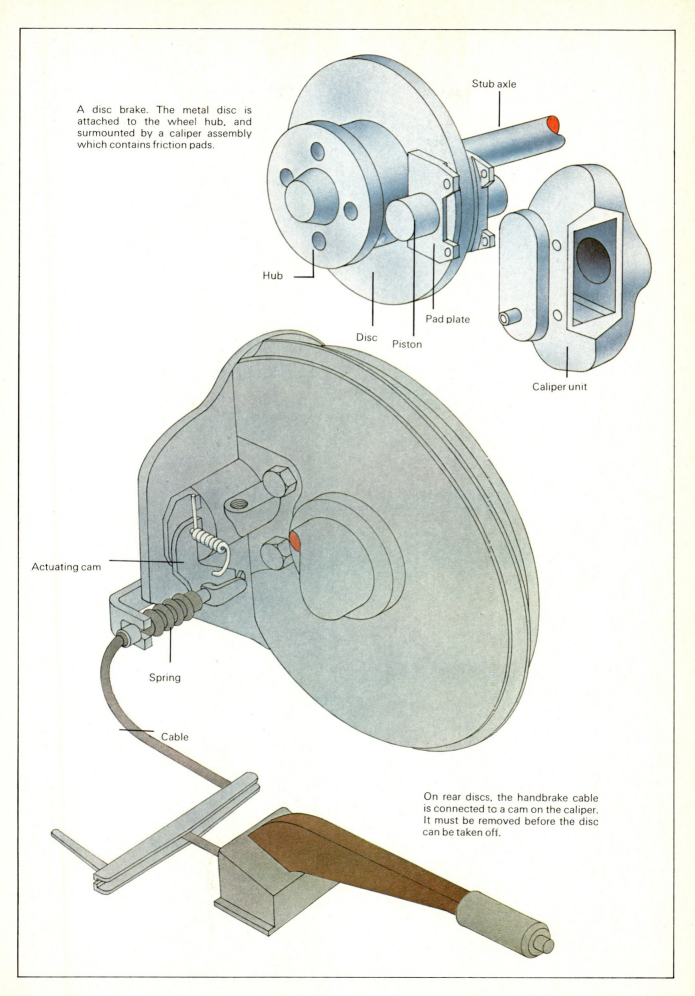

A disc brake. The metal disc is attached to the wheel hub, and surmounted by a caliper assembly which contains friction pads.

Stub axle
Hub
Disc
Piston
Pad plate
Caliper unit

Actuating cam
Spring
Cable

On rear discs, the handbrake cable is connected to a cam on the caliper. It must be removed before the disc can be taken off.

Servicing the handbrake

All cars are required by law to have an efficient handbrake system, both to hold the car when parked and to provide emergency braking if the hydraulic brake system fails.

The handbrake does not operate through the hydraulic system but usually through a series of either cables or rods, or a combination of both. Most handbrakes utilize the same brake linings or disc pads that are operated by the footbrake. In this case when the footbrake is adjusted at regular intervals it will automatically adjust the handbrake. However, after about 20,000 km (12,000 miles) the handbrake linkage will have stretched slightly and the travel of the handbrake lever in the car could be excessive, impairing the efficiency of the handbrake.

Separate handbrake pads
Some cars fitted with disc brakes have separate handbrake pads which have to be adjusted independently. Other cars, notably Volvos, are fitted with disc brakes all round, and have handbrakes which operate a small drum brake located inside the front or rear disc. Although there are a wide variety of handbrake layouts, designed to suit various footbrake installations, the general principles of checking and adjustment are very similar.

Service Checklist

9 Replace disc brake pads where applicable.

 10 Check and adjust handbrake.

11 Examine brake flexible hoses and metal pipes for security, chafing, leaks and rusting. Top up reservoir.

12 Check clutch hydraulics for rusting, chafing and leaks. Check clutch pedal travel, and adjust clutch clearance where specified.

Handbrake operation
The operation of the handbrake is initiated by the driver pulling on the interior handbrake lever. The lever may be mounted on the floor or on the dashboard but the function

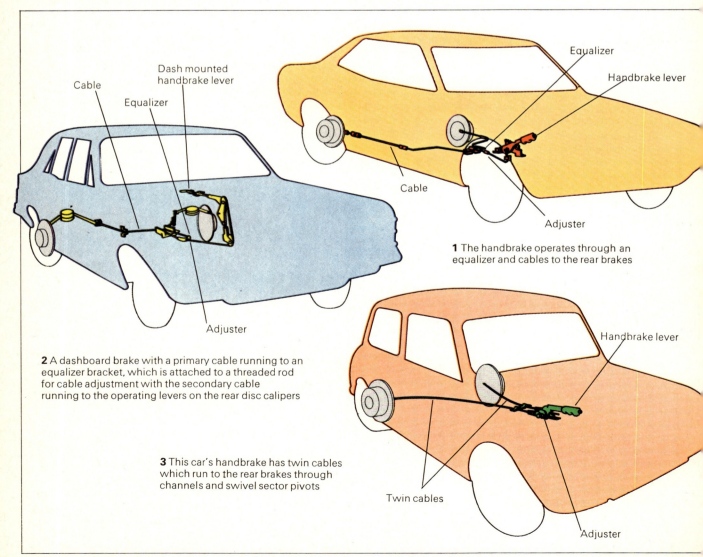

1 The handbrake operates through an equalizer and cables to the rear brakes

2 A dashboard brake with a primary cable running to an equalizer bracket, which is attached to a threaded rod for cable adjustment with the secondary cable running to the operating levers on the rear disc calipers

3 This car's handbrake has twin cables which run to the rear brakes through channels and swivel sector pivots

HANDBRAKE

is still the same and that is to act as a lever through which the pulling force from the driver is first amplified by the design of the lever and then transmitted to the road wheels via cables or rods. At the road wheels the cables or rods activate a lever mounted on the back plate or disc caliper which, as the handbrake is pulled on, will thrust out the brake linings or disc pads onto the brake drum or disc and stop the car from moving.

Preparations

Before attempting to tackle the adjustment of the cable or rod system, check that the rest of the inter-related brake system is working properly and efficiently. Where rear drum brakes are fitted, check that the brake linings have sufficient lining material remaining and that they are correctly adjusted (pages 61-65). If your car has disc brakes all round, check the condition of the rear disc pads (pages 70-72).

On any braking system where the handbrake uses the same brake linings or pads as the footbrake, the normal brake adjustment will take up any wear in the linings or pads, and will improve the operation of the footbrake and handbrake. Once you have determined that the footbrake is correctly adjusted, check the handbrake.

There are two main ways of checking the efficiency of the handbrake. The first is by road test. The car is driven along a flat straight road at 20 m.p.h. Making sure the road is clear both in front and behind, pull the handbrake on.

Handbrake linkages

These four cars illustrate the basic principles of handbrake control linkages

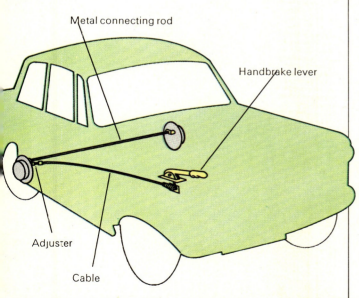

4 A single cable operates one rear brake with a connecting rod operating the opposite brake simultaneously

The legal requirement is that the car should be brought to a halt within 16 m (54ft.). For a rough check when the car is stationary, jack up the end of the car on which the handbrake operates and support the car on axle stands, making sure the vehicle is firmly chocked on the wheels which are still on the ground. With the handbrake off the wheels should turn freely, but when the handbrake is applied both wheels should be locked.

If the handbrake is not working efficiently, the fault will probably be in the mechanical linkage. This linkage could have become rusted and stiff or cables could be stretched and frayed, in which case adjustment or replacement will be necessary.

Types of handbrake linkage

The design of the handbrake linkage varies with nearly every make and model of car. However, certain design features enable the layouts to be categorized. Some models use twin cables to transmit the force from the handbrake lever to the wheels, while other models may use a combination of rods and cables. In some designs there are equalizing mechanisms to ensure that the force applied to each wheel is the same. A brief description of the main designs are listed below.

Twin Cable System: This is one of the simpler methods whereby the handbrake lever is attached to two cables, one for each wheel (fig. 3). There is no automatic balancing between the two cables, and this is achieved only by careful adjustment.

Rod and Roller System: In this layout a short rod with a roller at one end is attached to the bottom of the handbrake lever. Two cables pass around the roller, one for each wheel, in a loop arrangement having its fixing point at the brake drum or disc lever. The cable then passes around the roller and back to the brake backplate or caliper where it is attached by its adjusting mechanism.

Equalizer System: With this design a single cable runs from the handbrake lever to the equalizer mechanism. This mechanism is a pivoting arrangement which turns as the handbrake cable is pulled on. The pivot has two other fixing points from which rods run to each brake lever. Some cars may feature cables instead of rods (fig. 1).

Handbrake linkage adjustment

With all systems, jack up the wheels on which the handbrake works and support the car on axle stands. Before starting adjustment do a visual check of the linkage to determine the condition of the system and whether any replacement is necessary. You will need a helper to operate the handbrake lever while you watch that all the relevant mechanisms are moving freely. Cables should be cleaned of dirt and lubricated, and the wire and the outer casing should be in good condition. Some cables have a grease nipple in their outer casing and grease should be pumped until it is seen coming out of the end of the sheath (fig. 15). If no nipple is found it is possible to run thin oil down the cable from the highest end. This is done by dripping oil onto the inner cable and letting this run down into the outer sheath.

With other mechanisms check that rollers run freely and if a grease nipple is visible use a grease gun to lubricate the shaft. With an equalizer mechanism also check that the pivot turns freely and lubricate any moving parts.

Adjustment—twin cables: In this system each cable has to be adjusted separately by means of threaded adjusters, usually fitted at the base of the handbrake lever inside the car (fig. 6). Make sure that the handbrake is in the off position and then gradually tighten the adjusting nuts, first adjusting one a few turns and then the other. Each time you have tightened the adjusting nuts get out of the car and spin the wheel you are adjusting. When you hear the brake lining touching the drum as you spin the wheel, back the adjuster off slightly. The wheel should be still able to spin freely with the handbrake in the off position but when pulled on the wheels should now be locked.

Adjustment—single cables: The adjuster for single cables is usually under the car and positioned at the point where the cable meets the equalizer or roller mechanisms (fig. 7). To adjust the main cable identify the adjusting nut and lock nut. Then loosen the locking nut and tighten the adjusting nut until all the slack has been taken up. Again check that the wheels still spin freely when the handbrake is in the off position and are locked when the handbrake is pulled on.

Adjustment—rods: If there is a combination of rods and cables then it is the cables which normally carry the adjuster. On some older cars the rods carry adjusters, but the method of tightening is the same.

If the cable or cables have been stretched to the point where adjustment is no longer possible, new ones should be fitted (figs. 8 to 27). To do this, you will need a cable or cables made specifically for your car which can be bought from most dealers or good accessory shops. Under no circumstances fit incorrect cables to the braking system of your car.

With rear of the car jacked up and on axle stands, release the handbrake lever. Start removing the cable by working from the wheels forward to the handbrake lever. Discard any split-pins or bolts; these must be replaced on re-assembly.

Make sure all parts are moving freely and make sure you grease the cable before assembly.

Fit the new cable from the wheel end, adjust as described above.

The interior handbrake lever works on a spring loaded ratchet system. This method allows the handbrake to be locked on in a variety of positions. The main faults to be encountered are spring failure or a worn ratchet. In either case the only solution is replacement of the worn part which can be obtained from your dealer. Always check the condition of the ratchet because if this is worn it could cause the handbrake to disengage when the car is parked.

Careful application of the handbrake and regular lubrication can do much to reduce ratchet and pawl wear.

Where to adjust your handbrake

5 Rod and cable adjusters are situated by one rear brake drum, a connecting rod adjusts the other brake simultaneously

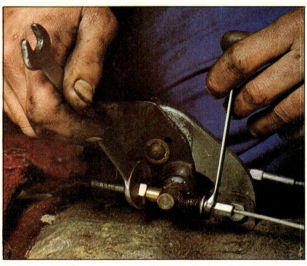

6 Twin cable adjusters are found either side of the handbrake lever. Each cable is adjusted separately

7 The equalizer system adjuster will be found where the front cable joins the equalizer

HANDBRAKE

Replacing a cable – twin-cable system

This step-by-step operation was carried out on a Morris Minor 1000

8 First you will have to undo the two nuts inside the car, beside the brake lever

9 Further back along the transmission tunnel find and release a retaining clamp, fastened each side of the tunnel

10 Situated below the seat in the corner of the foot well, is another bolt which must be undone

11 The other end of this bolt protrudes underneath the car, adjacent to the drive shaft

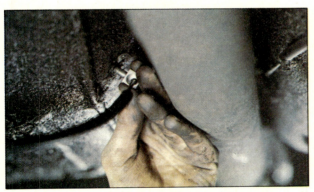

12 You will probably need help, but with the nut removed the bolt can then be pulled through

13 There is a nut which holds the outer cable to the wheel back plate, this should be removed

14 Extract the split pin, then pull out the clevis pin to separate cable from brake lever

15 The reverse procedure should be followed to fit the new cable, which should be greased with a suitable gun

Replacing a cable- equalizer system

This sequence was carried out on a Renault 12

16 First locate and release the nut which connects the equalizing unit to the front cable

17 Prise off the dust cover at the wheel hub and remove the split pin and holding nut

18 Remove the brake drum and, using a pair of pliers, remove the end of the brake cable from the lever

19 Feed cable through from the back of the brake drum, with the help of a screwdriver

20 Working now from both sides of the brake assembly manoeuvre the cable through the back plate

21 When you have freed the cable, push it through the lower radius arm

HANDBRAKE

22 Locate the cable joiner underneath the car and slide back the dirt excluder for access to the cable

23 The first step in dealing with the front end of the handbrake is to unhook the cable from the lever arm

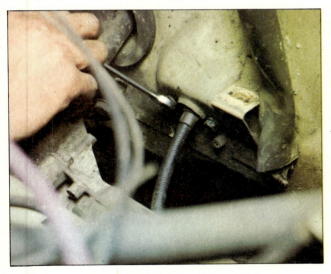

24 The release bolts which are located in the engine compartment must now be removed

25 Disconnect the battery and then unbolt the handbrake from under the dashboard

26 Draw out the handbrake arm and disconnect the electrical connection

27 Using a drift tap out the pin to release the cable. When re-assembling insert the pin from the opposite side

Checking brake hydraulics

Service Checklist

10 Check and adjust handbrake.

 11 Examine brake flexible hoses and metal pipes for security, chafing, leaks and rusting. Top up reservoir.

12 Check clutch hydraulics for rusting, chafing and leaks. Check clutch pedal travel, and adjust clutch clearance where specified.

13 Check cooling components for leaks and blockages. Check condition and tightness of fanbelt.

The hydraulic braking system uses fluid under pressure to operate friction brakes at the wheels. This pressure is generated in a 'master' cylinder and is transmitted through a network of steel tubes and flexible hoses to 'slave' cylinders at the road wheels, while a reservoir maintains a constant supply of fluid to the system. Fluid leakage which would result in pressure loss is prevented by rubber seals fitted to the master cylinder and to each of the wheel cylinders.

When the brake pedal is depressed, the master cylinder piston pressurizes the fluid in the pipelines to each wheel cylinder (known as the slave cylinders). This causes the wheel cylinder pistons to apply the thrust necessary for operating the brakes. Since the fluid pressure at each wheel cylinder when the brakes are functioning normally is equal, there is balanced braking on all four road wheels.

On some cars a 'vacuum servo' is fitted. This employs the vacuum produced by the engine to increase the pressure placed on the hydraulic fluid in the braking system. Where a vacuum servo forms a part of the braking system, it can either be a remote unit—inserted in the line between the master cylinder and the wheel braking assemblies—or a direct acting type mounted integrally with the master cylinder and connected directly to the brake pedal. Both British braking system manufacturers supply in accessory form a vacuum servo suitable for fitting to cars without a factory-fitted servo. In the case of Lockheed this is the 'Brakemaster' and in the case of Girling, the 'Power-stop'.

Additionally some cars are fitted with a dual line braking system which provides two independent hydraulic circuits, either of which will operate part of the vehicle's brakes in the event of a failure in the other part. The two circuits are supplied from a tandem master cylinder, which is often operated by a vacuum servo.

The hydraulic braking system will work efficiently and will not require bleeding as long as air does not enter the system and providing you change the fluid completely at periodic intervals, i.e. every 18 months. By doing so you ensure that there is no possibility of the fluid failing through a phenomenon known as 'vapour lock'. By its nature brake fluid absorbs moisture from the atmosphere. This added moisture in the system will boil under heavy braking stresses, when the friction causes extra heat build up, and if this happens total brake failure will result. Many garages now have a machine which can test a sample of brake fluid for moisture content.

Brake fluid

The brake fluid you use in the system must, at the very least, conform with SAE J 1703 and this specification must be shown on the can. These standards are important because brake fluid has to operate at extremely high temperatures produced by the braking friction, as well as at very low temperatures.

Braking fluid must also be compatible with the metallic parts and the rubber seals used in the braking system. Some models have employed a mineral based fluid, principally some Citroen models, and if you own one of these cars then you must ensure that you use only the correct type of fluid. Should the wrong type of fluid be used, drain the system and flush with methylated spirits. All rubber seals and cups should be replaced.

Bleeding the system

Removing air from the brake system is known as bleeding and the only equipment you need is a tin of brake fluid, a ring spanner for the bleedscrew on the rear of the brake backplate, a clean, see-through receptacle, a piece of translucent tubing and if possible one or two people to

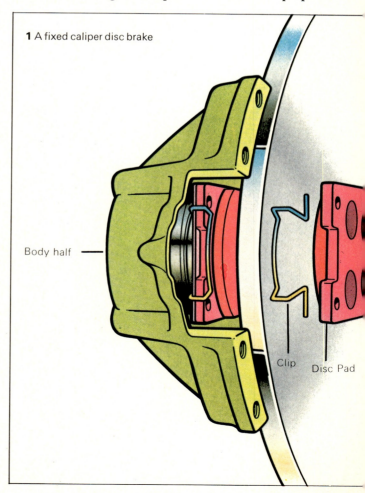

1 A fixed caliper disc brake

Body half — Clip — Disc Pad

BRAKE HYDRAULICS

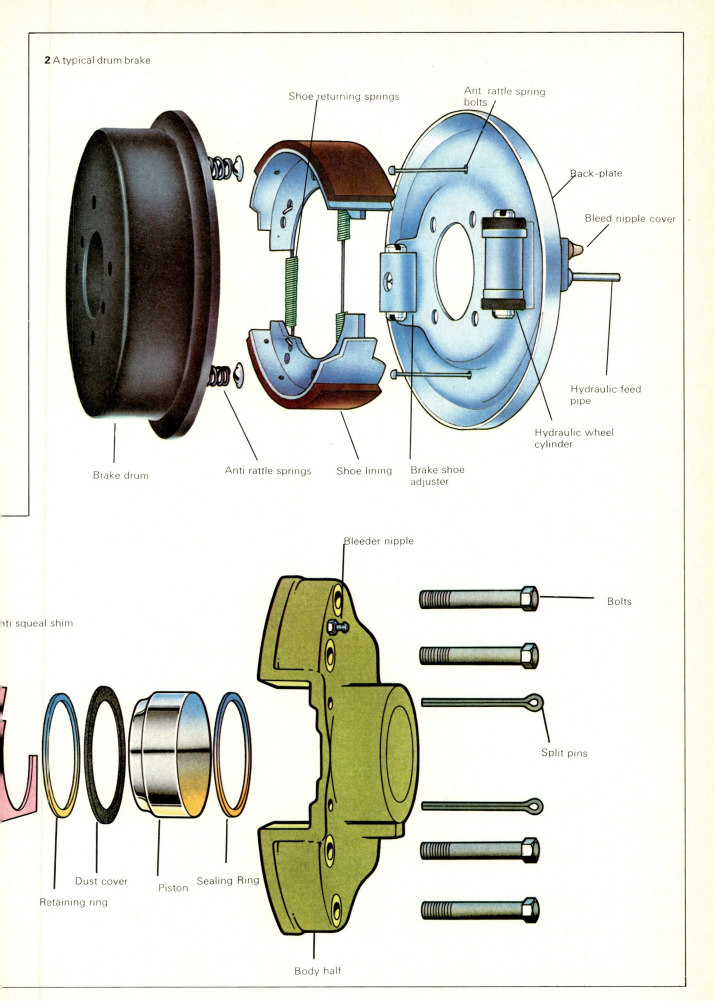

2 A typical drum brake

Bleeding the system

3 Fill the reservoir with brake fluid and keep it topped up throughout the operation to avoid taking air into the brake system

4 Fit the flexible tube to the cleaned bleed nipple with the other end in the fluid. Slacken the bleed nipple and pump brake pedal

5 Continue pumping until air is expelled from brake circuit. Tighten nipple and check that pedal is firm

assist you. Before you attempt to bleed the brakes, however, keep the following in mind:
1. The master cylinder reservoir must be kept topped-up with fluid at all times.
2. The action used to pump out the fluid varies according to the type of master cylinder fitted to your car.
3. Don't spill fluid on your car because it will ruin your paintwork.
4. Don't re-use fluid bled from the system. Dispose of it safely.
5. Remove carpets or anything else which obstructs the full stroke of the pedal.
6. Chock the wheels before releasing the handbrake. If you have a BLMC 1800 or a Peugeot then leave the handbrake on.

Before starting to bleed the brakes lift up the bonnet and find out which type of master cylinder is fitted to your car and then you will be able to use the correct pedal action (see panel, page 85).

Figs. 3-5 show how the job is carried out using only a tube and glass container. A helper will also be necessary. Automatic, single-handed bleed devices are also available.

Before you start bleeding clean the area around the top of the master cylinder filler cap. Then, taking care not to let any dirt enter the reservoir, remove the cap.

Fill the reservoir with brake fluid and keep it topped up with fresh fluid throughout the bleeding operation. If you do not, air will be drawn into the system and the operation will have to be begun afresh.

Taking each wheel in turn, and beginning with the one farthest from the master cylinder, attach the flexible translucent tube to the cleaned bleed screw of the wheel cylinder or the disc brake caliper and allow the other end of the tube to trail in a glass container containing enough fluid to well cover the end. (In the case of a tandem braking system, you may have to bleed both wheel units on one side of the vehicle at the same time.)

Using a suitable spanner, open the bleedscrew taking care not to damage the nut. Your helper should then depress the brake pedal. The pedal must be depressed, and the action repeated, until clear, bubble-free fluid emerges from the tube or tubes. Then with the pedal held fully depressed, tighten the bleed screw.

Repeat this procedure for a single brake line system for each of the remaining wheels or, for a dual system, at the brake assemblies for the wheels on the opposite side of the system, remembering to keep the reservoir topped up.

To finish, give the pedal a good pumping. It should feel firm and free from any spongy action, a condition indicative of air in the system. Then check that the fluid in the master cylinder is at the correct level, topping up if necessary and replacing the filler cap.

Before road testing, pump the pedal hard then hold it down applying maximum possible pressure for two minutes. At the end of the two minutes check all new and disturbed connections for leakage while the pedal is held down. Finally, taking extreme care if on public roads, road test the vehicle.

If the brake fluid is being changed completely then the procedure is the same as for bleeding the system, but you continue bleeding until the new fluid is seen in the translucent tube. When the pedal has been pumped twice more, it should be held and the bleedscrew closed. Each bleedscrew should be bled in turn but if air has not entered the system there is no need to go round each bleedscrew twice.

Drum brake wheel cylinders

Wheel cylinders and flexible hoses must be repaired or changed when they are either leaking or showing signs of failure. On some rear brakes there is a single piston cylinder fitted which must slide easily on the backplate if the brake is to work efficiently. Excessive wear on one shoe lining and an ineffective handbrake are the usual indications of the wheel cylinder not sliding.

With the car jacked up at the rear and then supported on axle stands, remove the rear wheel (fig. 7). Back off the brake

Brake Hydraulics

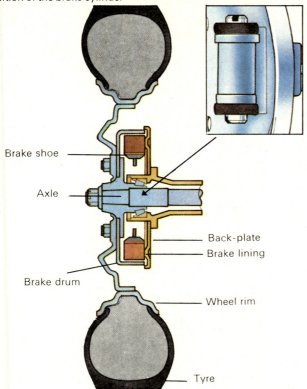

6 Section through a complete rear wheel assembly showing position of the brake cylinder

Checking drum brake wheel cylinders

7 Undo wheel nuts and remove wheel. With the other wheels safely chocked release the handbrake lever

8 Slacken the square-headed brake adjuster screw behind the brake plate so that the drum is completely free

9 Having marked positions of pull-off springs and shoes, release springs and remove shoes

10 Carefully clamp brake flexible hose with special clamp or a 'mole' wrench to reduce fluid loss from system

adjuster screw (fig. 8) with a brake adjuster spanner so that the linings will not foul the inside of the brake drum. Unscrew and remove the brake drum retaining screws and remove brake drum. If the drum is difficult to move, jar it by hitting the drum with a hide or copper hammer to free it on the hub studs. First look for signs of fluid around the brake cylinder. If any is present the seals at least will have to be replaced. The next check is to see if the piston is actually moving in the cylinder. On rear drum brakes, get your helper to push the brake pedal down very slightly and slowly. If the piston moves out, it should also be returned by the shoe springs. On no account pump the piston clear of its housing, as the system will require bleeding.

On front drum brakes, block the two brakes shoes one at a time with a piece of wood. The other piston should be moving normally when the pedal is slightly pressed. Again be careful with the piston travel.

If you decide to change the seals it will be necessary to buy an approved kit for your car. Don't skimp on time, and change the seals on the opposite wheel's cylinder as well.

Servicing a rear wheel cylinder

With the drum removed mark and note positions of the brake shoe pull-off springs and the brake shoes (fig. 9) this will ensure simple and correct re-assembly later. Release the pull-off springs from the web of the brake shoes and remove the shoes from the brake assembly.

As a further check to ensure correct re-assembly lay the parts down in their correct relative positions after thorough cleaning and critical inspection for condition as regards wear.

The next step is to ensure there is no unnecessary fluid loss. Either clamp the brake flexible hose with a special brake pipe clamp (fig. 10) or if without this tool clamp the flexible carefully with a 'mole' type wrench. An alternative method of preventing brake fluid loss is to seal

11 Undo brake pipe union nut and ease pipe away from back plate. Undo and remove the brake bleeding nipple

12 Remove circlip and washer from wheel cylinder boss showing through backplate. The cylinder is ready for removal

13 Remove wheel cylinder. It is vital that all parts are fully inspected and cleaned

14 Pull back the rubber dust cover from its groove in the outside end of the wheel cylinder. Repeat at the other end

15 Gently pull pistons out of the cylinder bore. Remove dust caps, taking care no fluid touches covers or bores

16 With all parts removed, check dust cover and pistons for wear and change lip seals

17 Inspect cylinder bore for wear. If it is worn it must be replaced. Otherwise clean with Trichlorethylene. No oil to be used

18 Re-assemble parts in reverse order. Check handbrake adjustment and thoroughly bleed the system

BRAKE HYDRAULICS

the entire system at the master cylinder by covering the top of the reservoir with a piece of polythene and then replacing the cap. To remove a wheel cylinder then you must first undo the brake pipe nut from behind the back plate (fig. 11). If you have a clean bleed screw rubber dust cap available, put this on the open end of the rigid brake pipe to keep the fluid in and the dirt out. Undo the brake bleed nipple with a ring spanner and remove.

On many vehicles the handbrake linkage lever must be disconnected from the cable at this stage. Remove the circlip and dished washer from the boss of the wheel cylinder protruding through the rear of the back plate (fig. 12). The wheel brake cylinder is now free for removal. Remove the wheel cylinder from the back plate (fig. 13) and place on a clean surface. Pull back the rubber dust cover from its groove on the outside end of the wheel cylinder (fig. 14). Withdraw the pistons from the cylinder bore (fig. 15). If the piston is seized it may be freed by soaking in methylated spirit, but it may well be more prudent to renew the cylinder. Lay all wheel cylinder parts out for inspection (fig. 16). Examine the dust covers, if there is any doubt regarding cracking or distortion renew the covers. Examine the condition of the pistons and change the lip seals (fig. 16). All seals should be changed every 60,000 km (40,000 miles) or at least every three years. To change the seal simply pull it off and replace with the new one making sure that the lip faces away from the piston. With kits, rubber grease is usually provided; but whether or not it is in

Hydraulic master cylinders

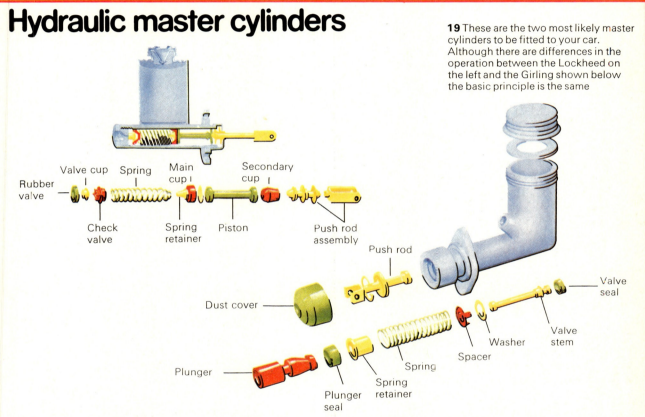

19 These are the two most likely master cylinders to be fitted to your car. Although there are differences in the operation between the Lockheed on the left and the Girling shown below the basic principle is the same

One of the most popular types of master cylinder is the Girling CV of which there are several designs. When dealing with this type, the pedal must be pushed down quickly through a full stroke, followed by three very quick half-strokes. The foot should then be lifted off the pedal so that it returns quickly to its stop. This pattern should be repeated until you are ready to tighten the bleed screw when the pedal should be held down on the last stroke.

The Girling CB master cylinder is employed mostly on the Rover 3 litre and light commercial vehicles. The pedal action is slow and deliberate. Push it slowly through its full stroke and return it slowly. Wait three or four seconds and repeat the pedal action.

When dealing with tandem master cylinders you must, of course, ensure that both master cylinder reservoirs are kept topped up. Push the pedal lightly down using a three-quarter stroke and return it slowly. Wait three or four seconds and repeat the action. Some cars have an electronic warning device fitted and during the bleeding operation a bulb will light on the dashboard. Ignore it until you have removed all the air from the system, then open a bleedscrew on a front brake and lightly press the pedal down. A check should be felt on the pedal and the light should go out. Close the bleedscrew immediately and release the pedal. If the light does not go out, open a bleedscrew on one of the rear brakes and try again.

Rear brake limiting valves (G-valves) are fitted to the BLMC 1800 and some Peugeot models and when bleeding, the handbrake should be left on. Bleed the front brakes using the pedal stroke necessary according to the type of master cylinder fitted. When bleeding the rear brakes, use the pedal action as for the CB master cylinder. If the pedal suddenly goes hard and cannot be pushed down any further, then the G-valve has been actuated. Release the foot pedal, close the bleedscrew and put the handbrake in the off position. This action will push the ball in the G-valve off its seat and let the fluid through. Re-apply the handbrake and continue bleeding.

If the system incorporates a vacuum servo, pump the brake pedal several times to destroy the vacuum in the servo unit and reservoir should one be fitted.

the kit you must only lubricate the seal with the specified rubber grease or fresh hydraulic fluid. When the piston assemblies are being replaced care must be taken to check that the lip seal is not turned back.

Looking at the replaced seal can give you a clue to the cause of failure. Straightforward wear on the lip is unavoidable, but if there are any signs of swelling or chemical destruction you can assume that the fluid is contaminated If so, the fluid must be changed, flushing the complete system and renewing all seals, including those in the master cylinder. The hoses may also have been damaged.

With the pistons removed, you may yet find you have to renew the whole cylinder if there is scoring on the bore surface. Inspect the bore thoroughly with sight and touch (fig. 17) for any imperfection that would require its replacement.

Re-assemble in the reverse order (fig. 18) after soaking all cylinder assembly parts in brake fluid.

Replacing a front wheel cylinder
Where two pistons are fitted to the backplate, as on front drum brakes, the procedure is similar, except that the wheel cylinders have to be unscrewed from the back of the brake backplate. Replacement is a reversal of the removal procedure, leaving the plastic plugs in the new wheel cylinder in place until you are ready to reconnect the pipes and hoses. At this stage screw in a new bleed screw, but take care not to overtighten it.

In some cases the wheel cylinder is designed to have a slight movement on the backplate (approx. 3 mm or .125 in.) and the fixing nuts should therefore be slackened half a turn or just enough to permit the cylinder to slide.

When the unit has been refitted and hoses and pipes reconnected bleed the system in the recommended manner. Check new and disturbed connections for leaks, refit the wheels, jack down and road test.

Disc brake hydraulics
Where disc brakes are fitted, the fluid travels down the pipes and flexible hoses in the same manner as for drum brake systems and then forces the piston in the disc brake caliper towards the disc. This fluid is kept in the system by seals either around the circumference of the piston or around the piston bores in the caliper. If any fluid can be seen to be leaking past these seals then new ones must be fitted.

To do this you must first jack up the car and remove the road wheel from the caliper on which you wish to work.

Remember to use axle stands. Remove the disc pads (see pages 67-68) and then gently press on the brake pedal to force the pistons slightly out of their bores. You can then place a piece of clean polythene over the master cylinder filler reservoir, after removing the cap, to prevent the loss of fluid. Retain the polythene by replacing the cap over it. Then disconnect the flexible brake pipe at the union with the rigid pipe which leads back to the master cylinder and plug the open end of this rigid pipe with a piece of clean plastic.

Unscrew the caliper mounting bolts and note any shims between the caliper and the stub axle lugs. Take great care not to tamper with the bolts which hold the two halves of the caliper together. In no circumstances must the caliper be dismantled.

Remove the piston dust covers (rear wheel discs only) and pull the pistons out of the caliper bores and carefully remove the piston seals and seal retaining rings. If the piston is difficult to shift, attempt to blow it free with an air-line or foot-pump attached to the fluid inlet port. Wash the caliper and piston assembly in clean brake fluid and check all components for wear and damage. In particular examine the outer surface of the pistons and if they show signs of scoring or rust damage then they should be replaced (figs. 31 to 34).

Clean out the grooves that secure the seals or retaining rings and wash the inside of the cylinder with clean fluid. Soak the new seals in clean brake fluid and fit them to the pistons or caliper cylinders. Lubricate the pistons with brake fluid and refit them in the caliper bores, ensuring that they are pressed in evenly all the way round. Retaining

21 Remove oil pressure light switch wiring from terminals on the underside of the tandem brake master cylinders

22 Remove the brake pipe from the outboard master cylinder. Carefully ease pipe union away from cylinder body

20 The flexible hose linking the hydraulic pipe to the wheel cylinders should be looked at regularly for corrosion and leaks

Brake Hydraulics

rings which fit flush with the end of the cylinders can be pressed into position with a brake pad and a piston compression tool.

Replace any dust covers which were fitted and make sure that the inner lip of the cover fits securely in the groove around the piston. Refit the caliper to the stub axle lugs and replace any shims. Fit the caliper bolts with a new tab locking washer. Tighten the bolts and bend the ends of the tabs over the bolt head. Lubricate the face of the piston with brake grease. Replace the pads and any shims, reconnect the pipes and bleed the system.

Pipes and hoses

Brake pipes and flexible hoses should be regularly checked for damage and corrosion. Bend the flexible pipes and look for cracks or signs of deterioration and chafing by wheels or suspension parts.

To replace a flexible hose at the road-wheel end of the system, hold the hexagon headed nut steady with a suitably sized spanner and undo the acorn nut on the metal pipe first. You can then remove the lock nut which holds the flexible hose to the mounting bracket. The other end of the hose can then be unscrewed from the caliper or wheel cylinder. If a washer is fitted then a new one should replace the old one. There are both UNF and metric threads on hoses and pipes. On metric pipes no sealing washer is used.

To refit the hose you should first screw the unit into the wheel cylinder and ensure it is fully tightened. Then thread the pipe end through the bracket and secure it with the lock nut. Before tightening the lock nut ensure that the pipe is not twisted or likely to come into contact with the wheels or suspension. Refit the rigid pipe to the end of the flexible pipe.

When replacing rigid metal piping you must ensure that you have an identical replacement piece of piping, in both thickness, length and shape. Custom cut lengths of pipe, with the correct flaring at the ends are available, but if you have to have the pipe made up by a specialist, make sure that the flaring is correct. If the pipe is to fit into a flexible hose it should have a 'female' flare and if it is to fit in a master cylinder, wheel cylinder or three way connector, then it must have a 'male' flare.

When threading the pipe into its position under the car, tape up the ends to prevent dirt from entering and then check that it fits without distortion. Connect the pipe at both ends ensuring that you tighten the union nuts correctly. Then bleed the brakes and road test.

Replacing master cylinders

Master cylinders, like wheel cylinders, require no routine maintenance but brake manufacturers recommend that they should be replaced every 60,000 km. (40,000 miles) or three years to ensure continued working safety. If fluid is leaking from the master cylinder then it must be replaced immediately. Before removing it, wipe the master cylinder clean with a piece of rag. Also clean the pipe nuts and pipe connected to the master cylinder. Compare the new master cylinder with the old one. The bore dimension is usually cast on the side of the reservoir or it may have a small tag attached to it, and in this way you can tell if you have the correct replacement unit. Do not be tempted to fit a unit about which you have doubts—it must be right.

23 Remove the brake pipe from the inboard brake master cylinder. Take care to protect paintwork from fluid dripping

24 Slacken and remove the nuts securing the tandem master cylinder reservoir unit

25 Remove the unit keeping your fingers over the brake pipe entry holes to prevent dripping brake fluid, then drain fluid

26 Re-assemble the tandem master cylinder. Fill reservoirs with brake fluid and thoroughly bleed system

Remove the filler cap from the master cylinder, place a clean piece of plastic over the fluid reservoir and screw the filler cap back on. This creates a vacuum and prevents the loss of most of the fluid from the reservoir, and, as the system has not been drained, time and fluid will be saved later. If your car is fitted with a mechanical servo or brake booster the job is much easier because it is not necessary to disconnect any pedal linkage. Slacken the pipe nuts and unscrew the securing nuts. Ease the pipes away from the cylinder sufficiently for the master cylinder to be removed. Note the position of the seal between the servo and the master cylinder. Use a piece of rag to wipe the front face of the servo clean.

If your car is fitted with a Servo assisted tandem master cylinder, with pressure lights proceed as follows to remove master cylinder.

Remove oil pressure light switch wiring (fig. 21). Remove brake pipes from outer master cylinder (fig. 22) and repeat for inner master cylinder (fig. 23).

Remove nuts securing master cylinder assembly to the servo unit (fig. 24) and finally remove the master cylinder unit from your car (fig. 25). Re-assemble in reverse order (fig. 26). This sequence was carried out on a Toyota Celica which has a typical tandem master cylinder.

If your car has no servo then the job is a little more complicated. From the inside of the car you must remove the pin which holds the master cylinder push rod to the brake pedal. Replace the pin if it is broken or badly bent. Move back to the master cylinder, slacken the pipe nuts and remove the nuts securing the master cylinder to the bulkhead and put them carefully to one side. Carefully unscrew the pipe nuts and ease away from the cylinder to allow their removal. Take hold of the master cylinder in such a manner as to cover the ports with the fingers and prevent fluid from pouring out and remove the master cylinder from the vehicle. Note the position of any spacers, distance pieces or nylon inserts that may be fitted. These must be replaced with the new master cylinder so make sure you know where they came from. Wipe the bulkhead clean.

Check again that you have the correct master cylinder and in particular check that if it has a push rod that the replacement master cylinder has one of the exact same length. Some push rods have adjustable clevises; if this is the case adjust the length of the new one until it is the same length as the old. Check the brake pipe for corrosion, signs of fracture or damage, paying particular attention to the pipe nut which should screw easily into the master cylinder port. It it doesn't, then don't force it because something is wrong. Once again the threads could be either UNF or metric and you must have the correct type for your car. The threads may be damaged or the end of the flare could be out of shape. Provided the threads are the same, the safest thing to do is to repeat the thread test with a new pipe and nut. If this screws in easily then a new pipe and nut are the answer and these must be fitted. Move the foot pedal up and down by hand and check for freedom of movement. The bushes may require lubrication or the pedal return spring may be broken. If the action is not smooth over the range of travel, discover the reason and put it right.

If a servo is fitted there is no need to alter the height of the rod which operates the master cylinder. This is correctly

Dismantling a fixed caliper disc brake unit

27 Jack up front of car. Support car on stands. Clamp the flexible brake pipe with special clamp to minimise fluid loss

28 Undo the flexible hose to the front disc brake caliper. Take care to avoid any fluid falling onto painted surface

29 Remove the four clips holding the cylinder guides by pulling them away with a pair of pliers

30 Prise out the cylinder guides with a small screwdriver. Lay out clips and guides in order to help when re-assembling

BRAKE HYDRAULICS

set and should never be disturbed. If the rod is damaged in any way the only answer is a new servo. Fit the new master cylinder into position and refit the nuts and washers. Re-connect the pipe, but do not use excessive force on the pipe nut.

If the car has no servo, refit the spacers, distance pieces or nylon inserts back in their original positions and fit the new master cylinder to the bulkhead. Secure with nuts and washers. Re-connect the pipes but take care not to over-tighten them.

Reverse the dismantling procedure and re-connect the master cylinder push rod to the brake pedal. If adjustment is necessary, make sure that the brake pedal is kept in the fully back or off position and the master cylinder push rod is pulled back and held against the washer and circlip which act as stops. Then adjust until the pin can be pushed into position without strain. If there is a lock nut tighten it in position.

On all master cylinders there should be a minimum of 1 mm (.04in.) free play before any resistance is felt. Move the push rod in and out to check that this free play is evident and reset the push rod adjustment if necessary.

Bleed the system, but before road testing, pump the pedal hard and hold it down for two minutes. With the pedal held down check that there is no fluid leaking from any of the joints.

Checking the servo

Dealing with a faulty servo unit can be a tricky job if only because they are very prone to dirt contamination. Therefore, be very sure that the fault does not lie in another part of the system. As all servos are powered by the engine, a good check is to remove and block the tube from the engine (usually attached to the inlet manifold) and try the brakes under controlled driving conditions. If no difference is noticed then the fault lies with the servo. If a much greater effort is required at the brake pedal, then the servo is functioning properly.

There are two basic types of servo: piston and diaphragm. These are very different in shape, but both usually have a replaceable air filter which should be changed every 60,000km (40,000 miles).

With the piston type servo, all pistons, piston rods and cylinders should be checked for wear and corrosion. With the other type of servo, a new diaphragm may be necessary. In both cases, always work in very clean conditions, with clean hands. Use only rag that will not fragment to clean and replace all seals. Only specified brake lubricant should be used where necessary. If you are in any doubt about this operation, a replacement unit would probably be a better idea than overhaul.

31 The caliper can now be removed from the disc brake assembly. Inspect for signs of leakage

32 Gently prise the cylinder's piston dust cover out of the housing taking care not to damage the piston

33 After removing the flexible pipe drain the fluid into a suitable container

34 If the piston is difficult to pull out of its bore apply pressure with an airline to the fluid inlet port

35 After servicing the caliper, re-assemble all parts in reverse order. Finally rejoin the flexible hose to the rigid brake pipe

Servicing the clutch

Service Checklist

11 Examine brake flexible hoses and metal pipes for security, chafing, leaks and rusting. Top up reservoir.

★ 12 Check clutch hydraulics for rusting, chafing and leaks. Check clutch pedal travel, and adjust clutch clearance where specified.

13 Check cooling components for leaks and blockages. Check condition and tightness of fanbelt.

Most hydraulically operated clutches have spring-loaded slave cylinders which are self-adjusting. As wear takes place in the clutch, the push-rod, positioned between the clutch slave cylinder and the end of the clutch operating fork, will move outwards, automatically taking up any free play.

Checking hydraulic clutches
The piston, inside the slave cylinder, slides forward. The space left inside the slave cylinder barrel is filled with hydraulic fluid, thus automatically keeping the free play at the required tolerance. Hydraulic fluid for the slave cylinder is kept in a reservoir attached to the master cylinder. The master cylinder is located in the engine compartment of the car and connected by a rod to the clutch pedal.

Other hydraulic clutches have a method of adjusting the length of the push-rod. First identify the type of system fitted to your car. Look underneath and locate the slave cylinder, usually attached to the clutch bell-housing and connected by a rod to the operating fork. If the push-rod is threaded, then the clutch is adjustable.

Adjusting the clutch
Disconnect the clutch pull-off spring, positioned between the operating fork and the clutch bell-housing (fig. 9). Check the amount of free movement in the clutch operating fork. It should be no less than 3 mm ($\frac{1}{8}$in.). If the play is excessive or there is no play at all, then the clutch needs adjusting.

Two spanners are needed to adjust the clutch, one to hold the push-rod and the other to loosen the lock nut. Hold the push-rod firmly with one spanner and free the lock nut with the other, one full turn of the nut is sufficient if the rod is to be lengthened, more if the rod is to be shortened.

To reduce the free play clearance screw the rod out. To increase the clearance screw the rod in. When the correct free play has been achieved, tighten the lock nut (fig. 14)—check that the clearance is correct and re-adjust if necessary—and reconnect the clutch fork pull-off spring. Depress and release the clutch pedal a few times and re-check the

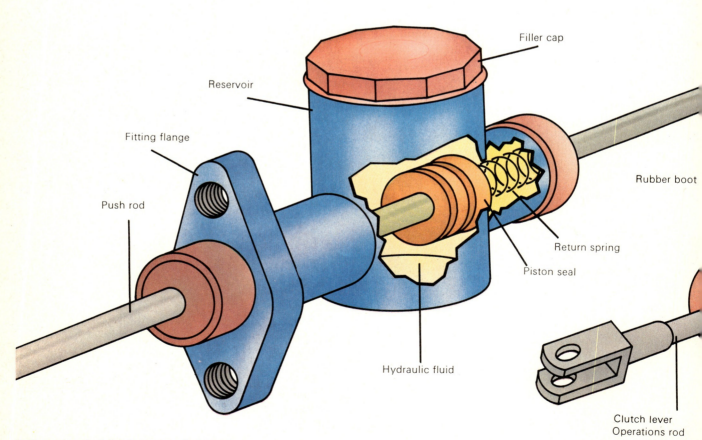

1 *Above:* A typical clutch master cylinder and (*above right*) the clutch slave cylinder

CLUTCH

clutch fork clearance.

One type of hydraulic clutch that can be adjusted, for it differs from other hydraulic clutches, is that fitted to the Leyland transverse engined models.

In this case, the adjustment is made at the clutch end of the engine (off-side of the car) from under the bonnet. Disconnect the pull-off spring between the top of the clutch operating arm and the back of the clutch slave cylinder, which is bolted to the top of the clutch housing. Pull the arm away from the clutch housing and, using feeler gauges, measure the distance between the adjustable stop screw and the arm. The correct clearance is .5 mm (.020in.).

To adjust the clearance, loosen the lock nut and turn the adjusting screw anti-clockwise to reduce the clearance and clockwise to increase it. When correct, it should just start to 'pinch' the feeler gauge. Once correct, tighten the lock nut and refit the pull-off spring.

It is essential to ensure that the hydraulic fluid reservoir is kept topped up at all times. Any adjustment will invariably involve fluid displacement, which must be replenished.

Topping-up the clutch reservoir

This is not quite such a simple job as may be implied, and certain precautions are very necessary.

First carefully wipe clean the top cap and body of the reservoir. Remove the cap and lay it on a piece of clean cloth. Avoid any contamination at all costs.

Avoid shaking the can of replacement hydraulic fluid. Position the top of the can as close to the reservoir filler neck as possible. Pour the hydraulic fluid in very carefully and slowly (fig. 13) until it reaches the correct level.

If any fluid is spilt the area should be wiped off and washed with cold water, particularly on paintwork.

2 To remove a master cylinder to check the seals, undo the nut securing the hydraulic pipe to the master cylinder

3 Remove pin connecting push rod to clutch pedal then undo bolts holding clutch master cylinder to bulkhead

4 The master cylinder can now be taken out. If the reservoir is attached to it, pour away the hydraulic fluid

5 Peel back the end cover and use long nosed pliers to squeeze together circlip so that it can be removed

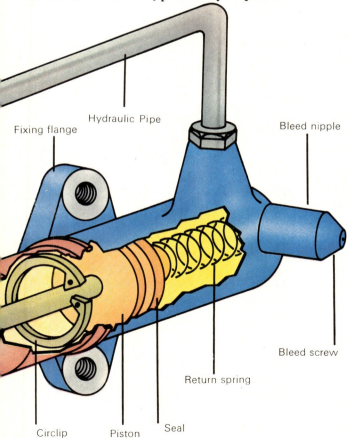

6 After removing the push rod, withdraw the piston assembly. If it sticks apply gentle air pressure to the reservoir hole

7 Having carefully inspected the cylinder bore for wear, replace the seals indicated before reassembly. Never use old seals

8 After reassembling and refitting the master cylinder, work under the car. Loosen the hydraulic fluid pipe joint

9 Using pliers unhook the return spring between the clutch fork and bell-housing. Take care it is a powerful spring

10 Undo the fixing bolts and remove the slave cylinder from the bell housing. Clean the mounting area and the bolt threads

11 Rotate the cylinder to remove it from the hose taking care to catch any loss of hydraulic fluid. Cover or seal clutch pipe

Check that the air-vent hole in the top of the reservoir cap is clear and that the cap is clean, both inside and outside. Refit the cap to the reservoir filler neck and tighten down by hand.

Checking hydraulic pipework

There are both steel tubing and flexible pipework on clutch hydraulic systems. The steel tube runs from the master cylinder to the bottom of the engine compartment scuttle panel, terminating at a union bracket. The flexible hose runs from the bracket into the slave cylinder. This allows for movement of the engine and transmission system without causing fractures of the pipework. Regular checks on both sets of pipe are essential.

Flexible hoses, which may be rubber or plastic, should be regularly checked every few months for signs of chafing or perishing. Get a friend to press the clutch pedal inside the car and check to see if the hose swells up slightly at any point. This would indicate a weak area in the hose. At the same time check all union joints for signs of 'weeping'.

If a joint is leaking, tighten it. If the leak persists renew the hose, making sure new sealing washers are used at the joints.

On the metal pipework check for signs of rusting, this may be particularly evident at points where the pipe is secured to the bodywork with metal clips. Check also for kinking of the pipework and possible damage that could have been caused by flying stones. If any of these faults are found, the pipe must be renewed, no matter how minor the damage may appear.

Renewing clutch pipework

Before undoing any of the pipes, first drain the hydraulic fluid from the system. Loosen the bleed nipple on the side of the slave cylinder, fit a tube on the end of the nipple and place the other end of the tube in a jar, pump the clutch pedal to force out the hydraulic fluid. Never just undo a union and let the fluid run out. If it contacts the paintwork

CLUTCH

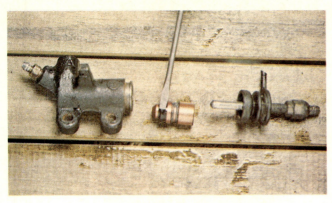

12 Again, carefully inspect the cylinder bore for wear, and replace the seals indicated before reassembly. Do not re-use old seals

13 Reverse sequence to reassemble, tighten all unions securely. Refill the reservoir with clean hydraulic fluid and bleed

14 If the clutch is adjustable, slacken off the nut with a spanner to give a clearance of 3mm ($\frac{1}{8}$in). Retighten the locknut

of the car it could cause severe damage to the surface.

To renew a steel pipe, undo the union where the pipe is connected to the flexible hose. Release the nut holding the pipe to the master cylinder and either undo or bend back the retaining clips. Lift the pipe off.

Buy the correct replacement pipe. This will have accurate bends already in it. Do not attempt to bend pipes without the proper tool. If a pipe needs shaping take the new and old pipe to your local garage and get them to bend it for you.

Refitting the pipe is just a reversal of the dismantling procedure. But do make sure the unions have washers on them, where applicable.

Flexible pipes are held to the union bracket with a lock nut. Undo the steel pipe from the union first. Undo the lock nut and then pull the flexible pipe away from the union bracket.

When replacing the flexible pipe, first screw it into the slave cylinder, making sure new sealing washers are used.

Slide the threaded end of the pipe through the union bracket and fit the lock nut. Before fully tightening the lock nut make sure the pipe does not have any kinks or strains in it. Finally connect the steel pipe to the end of the flexible hose.

Bleeding the clutch
Although the basic principle of bleeding all hydraulic systems is the same i.e. to remove any air from the system, the actual sequence used may vary in detail. The most likely hydraulic clutch systems are the Girling and Lockheed.

In both cases, however, the preparation is the same. Wipe the reservoir cap clean and remove it from the reservoir body. From underneath the car fit a bleed spanner to the bleed nipple (fig. 15), located at the side of the clutch slave cylinder. Attach the bleed tube to the nipple (fig. 16) and immerse the other end of the tube in a clean jar filled with about 25 mm (1in.) of new hydraulic fluid (fig. 17). Make sure the end of the tube is kept immersed in the fluid.

Bleeding a Girling system
Top up the reservoir with new clean fluid recommended by the manufacturers. From underneath the car loosen the bleed nipple by turning it anti-clockwise. Get a helper to press the clutch down slowly, making sure the helper keeps the reservoir topped up with new hydraulic fluid during the bleed operation. Never re-use the old fluid, it will be contaminated and contain a proportion of air.

Get your helper to keep pumping the pedal until there is no sign of air being emitted from the end of the bleed tube. The pedal should finally be depressed slowly and as your helper does so, tighten the bleed nipple by turning it clockwise. On this last stroke the helper should hold the clutch pedal down until the nipple is fully tightened.

Top up the reservoir, replace the cap, remove the bleed tube from the end of the slave cylinder nipple. Check the action of the clutch to ensure it is working correctly.

Bleeding the Lockheed system
The same basic principle applies as described for the Girling system. The difference, however, is in the actual pumping action. As the pedal is depressed, and just before it reaches the end of its travel, the bleed nipple should be locked off. Allow the clutch pedal to return under is own power, loosen the bleed nipple and depress the pedal, locking off the nipple as the pedal reaches the end of its travel.

This operation is repeated until the air is expelled from the system.

Checking mechanical clutches
Most cable or rod actuated clutches are equipped with an adjuster at the clutch operating fork end of the system. There are two basic methods of checking the adjustment, at the fork end or at the pedal. Clutch clearances do vary, but generally speaking an 3 mm ($\frac{1}{8}$in.) free play at the clutch operating fork is sufficient.

Where adjustment is made at the pedal, there should be approximately 12.5 mm ($\frac{1}{2}$in.)—25 mm (1in.) free play at the pedal before the clutch release bearing is felt to take up.

Adjusting mechanical linkages
First check in the owner's handbook to determine the actual clearance. This can vary from between 3 to 6 mm ($\frac{1}{8}$in. to $\frac{1}{4}$in.) at the clutch fork, depending on the make and model of car.

First remove the clutch pull-off spring and check the free play at the fork. Loosen the lock nut on the adjuster screw by holding the adjuster nut firmly with one spanner

15 To bleed clutch, first remove reservoir cap, then locate the slave cylinder bleed nipple and slacken it with a spanner

16 Attach a bleed tube to the nipple and immerse the other end of the tube into a glass jar partly filled with hydraulic fluid

17 Keeping the master reservoir topped up, get a helper to pump the clutch pedal until no air emerges through bleed tube. Finally tighten the nipple

while the lock nut is undone with another (fig. 18).

If there is too much free play, shorten the adjuster by turning the adjusting nut in a clockwise direction. If there is insufficient free play turn the adjusting nut anti-clockwise. This system applies to both cable and rod linkages.

Should the handbook specify free play at the pedal, then the following procedure should be adopted. Hold a ruler against the toe-board inside the car alongside the clutch pedal. Press the pedal down until the initial movement is taken up. This is the point where greater effort is needed to move the pedal further. It is this initial movement that is the free-play. Measure the distance the pedal travels (fig. 19).

If this is not in keeping with the distance specified it is simply adjusted at the clutch fork in the manner described previously. Turn the adjusting nut clockwise to decrease pedal travel and anti-clockwise to increase it.

The exceptions

The two exceptions to the common method of adjusting a mechanical linkage is the Ford Cortina Mk III and the Escort series.

On the Ford Cortina detach the release lever gaiter on the clutch housing and smear a small amount of molybdenum base grease on the ball end of the release cable. Refit the gaiter. Loosen the lock-nut on the adjuster underneath the car (fig. 21) and get a helper to press the clutch pedal to the floor. Pull the adjuster away from its housing and measure the distance between the adjuster nut and the housing with feeler gauges. The correct clearance should be 3.048 mm (0.12in.). Turn the adjusting nut in to reduce the clearance or out to increase it. When the correct clearance has been obtained, hold the adjuster firmly and tighten the lock-nut (fig. 22).

The Ford Escort range of clutch has no evident free-play at the clutch fork. The clutch pedal should be level with the brake pedal. If the clutch pedal is lower than the brake pedal, loosen the lock-nut on the adjuster screw under the car at the clutch fork. Turn the adjuster screw in, until both pedals are level. Hold the adjuster firmly and tighten the lock-nut. If the pedal is too high, reverse this procedure.

Lubricating linkages

If the clutch pedal is stiff, check and lubricate all the pivot points. This will include the shaft and the pedal pivots and, on rod linkages, any other joints through the system.

Adjusting a mechanical linkage on a Vauxhall Viva

18 To adjust freeplay of a clutch fork, usually 3-6mm, slacken adjuster nut until correct distance is found, then tighten

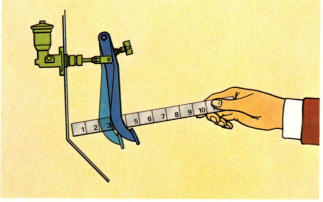

19 Diagram showing how to check pedal travel before clutch release bearing is operated. 25mm (1 in.) is an average distance

CLUTCH

If a cable operated clutch is stiff, lubricate the cable. This entails taking the cable off the car. First undo the locknut on the adjuster at the clutch fork and screw the adjuster in to slacken the cable. Disconnect the cable from the clutch fork (fig. 24) and unscrew the adjuster from its housing. Draw the cable away from the clutch fork, disconnect the cable from the clutch pedal, usually attached by a nipple on the cable fitting into a socket on the pedal (fig. 20).

Hold the cable up and trickle oil down between the inner cable and outer casing. Move the inner cable up and down inside its outer casing, to ensure it is free (fig. 26). Refit the cable to the car and adjust the clutch to its correct clearance.

Regular maintenance and attention to the clutch will ensure long clutch life and eliminate sudden failures.

Adjustment or removal of clutch cable on Cortina Mk III

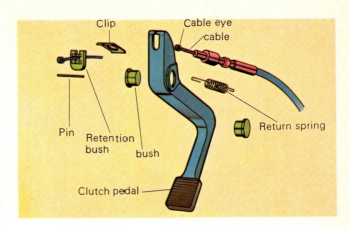
20 An exploded view of the component parts of a Ford Cortina MK III clutch operating pedal assembly

21 Slacken the locknut until there is a clearance of 3.048mm between the adjuster nut and the housing

22 Hold the adjuster nut firmly and tighten the locknut ensuring that the gap remains the same.

23 If the cable has to be removed, slacken the adjuster nut and ease the cable away from the clutch housing

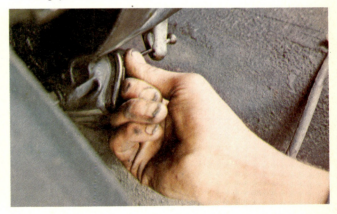
24 Free the ball end of cable from clutch fork take care not to dislodge the release bearing from clutch lever

25 Remove the rubber or nylon disc from the face of the bulkhead. Then disconnect the cable from the clutch pedal

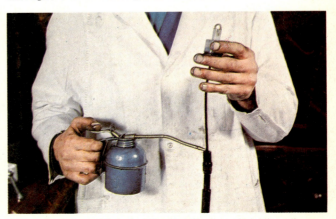
26 Lubricate the cable by holding it vertically and trickling oil in at the top

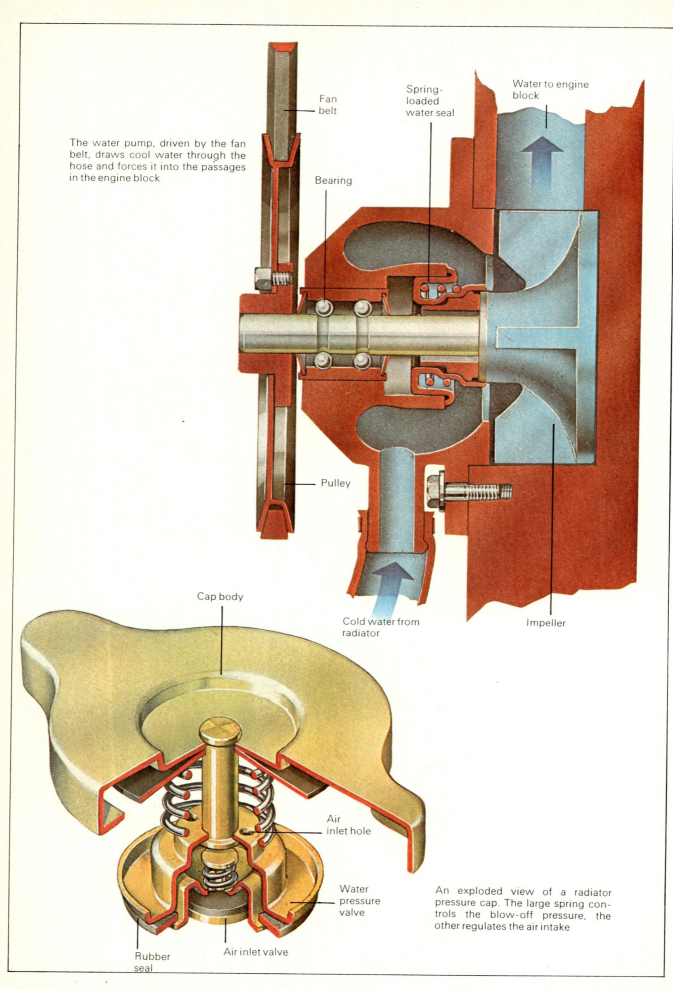

The water pump, driven by the fan belt, draws cool water through the hose and forces it into the passages in the engine block

An exploded view of a radiator pressure cap. The large spring controls the blow-off pressure, the other regulates the air intake

COOLING SYSTEM

Checking the cooling system

Service Checklist

12 Check clutch hydraulics for rusting, chafing and leaks. Check clutch pedal travel, and adjust clutch clearance where specified.

 13 Check cooling components for leaks and blockages. Check condition and tightness of fanbelt.

14 Check operation and condition of all electrical equipment. Adjust headlamp settings if required.

The cooling system is one of those sections of the car that tends to be taken for granted. A system in sound order needs only occasional topping-up and the only times when it makes its presence felt is when the heater fails or a hose splits in a welter of steam.

A fault in the cooling system can result in, at best, a long delay while repairs are made and, at worst, a totally ruined engine. Less spectacularly, a defective cooling system will increase petrol consumption and engine wear. Other effects can include localised overheating causing pre-ignition, running-on, pinking and even distortion of the cylinder head.

The system should be carefully checked at the recommended service intervals and an extra inspection carried out before a long trip.

Progress in engine design has resulted in cooling systems becoming physically smaller over the years and working much more efficiently. This also means, however, that

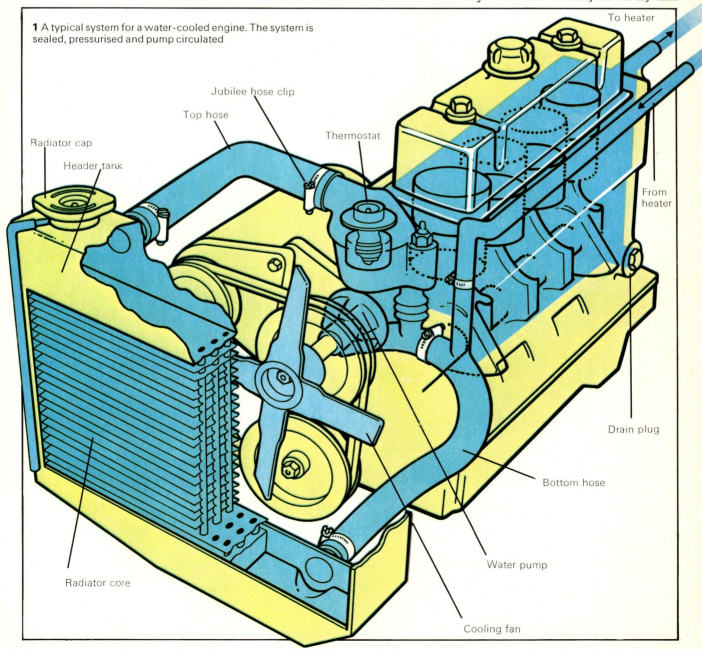

1 A typical system for a water-cooled engine. The system is sealed, pressurised and pump circulated

2 Basic components of the cooling system which are most likely to need replacement at regular intervals including fan belt, various hoses, thermostat, radiator cap and Jubilee clips

the system is more highly stressed than it was a generation ago and is, as a result, more vulnerable.

Check the system regularly

Given a correct checking routine with usually low cost replacements being fitted when necessary, there is no reason why the cooling system should ever give trouble. A combined distance/time formula is the guarantee against trouble. If a figure of about 5,000 km (3,000 miles) is accepted as the distance interval, potential troubles can be nipped in the bud. The distance period should be combined with a time-interval to be sure of perfect service. Once a month is as good a time interval as any with an extra check before a holiday trip or extra-long run.

A further time-interval to be taken into consideration is the lifespan of the fan belt and hoses. A fan belt is suspect after one year and hoses—especially the top hose—have a certain life of two years. After these times which are, perhaps, on the cautious side, the components are suspect. They should be examined with extra care and if there is doubt about their condition, they should be renewed.

All this background detail is needed if a proper check of the cooling system is to be made—a process which does not take long and is one of the few which does not call for special tools or instruments. It is, in fact, largely a visual routine because many incipient faults can be fairly easily seen. A leak is often betrayed by the streaks left by coloured antifreeze or the rusty stains of corosion. A decaying hose can also be easily seen in many cases, so a close visual check forms a large part of maintenance.

Although engine layouts vary, the cooling system arrangements all follow a similar pattern starting at the radiator and working back along the line.

It is best to start a check with a cold engine so that the radiator caps can be removed and clips undone readily. A hot engine will have hot water in the system and the water will be under pressure.

The radiator cap

The radiator is a good place to start the check and there are several points to watch for.

Inspect the area round the filler cap first to make sure the overflow pipe is sound. It may have corroded near the point at which it leaves the filler stub or further along its run. While a fault in the pipe may not affect the operation of the system itself, a dribble of liquid oozing out from high up on the radiator could be blown back into the engine compartment and settle on plug leads or other ignition components to cause tracking and consequent misfiring.

The radiator filler cap seals the cooling system to the set pressure. If the pressure rises above the predetermined level, a valve in the cap lifts off its seating to allow the excess pressure to escape through the overflow pipe. A non-return valve arrangement in the cap allows air to be drawn into the system when the engine cools down and the volume of water contracts.

The cap condition is best checked using a pressure tester, but if one is not available a visual check can be made of the cap mechanism. A broken or distorted spring cannot be missed and a defective valve is equally obvious by merit of the ragged condition of the perimeter. A further check can be carried out by starting the car and letting the engine warm up for a few minutes. Using a cloth as protection against hot water, carefully undo the cap for about one half turn. A slight hiss should be heard as the pressurised air is released. If no hiss is heard, check the rubber on the sealing ring and if this is damaged in any way, fit a new cap.

It is not possible to repair a cap—a replacement has to be bought but it is most important to fit the correct cap. The pressure-rating may be stamped on the top of the cap and the replacement must be the same rating.

The higher the pressure, the higher the temperature at which the water boils, so fitting a cap of a too low pressure rating could result in overheating and general loss of efficiency. Too high a rated cap could result in joints blowing in the cooling system.

When the engine is cold, the water level in the radiator may appear to be low, especially if a high-pressure system is used. This low level may be due to the fact that expansion room accounts for the gap. In some designs there may be some form of level indicator in the shape of a metal tongue or a mark on the inside of the filler orifice.

If the level is below any such indicator, it may be due to nothing more alarming than a slight loss through evaporation or other allowable factors, or it could be the result of a leak just starting to make its presence felt. Do not worry about it at this stage, but bring the level up to the indicated mark, or just below the overflow outlet and make a mental note to check the level after a run in the car.

Radiator leaks

The radiator unit as a whole needs careful checking. Look closely at the joint between the header-tank unit at the top and the main body of the radiator. A weeping joint will be made fairly obvious by stains of coloured antifreeze or rust.

Such a fault is serious because it will inevitably become worse so treatment is fairly urgently needed.

It may not be enough just to repair the leak—the radiator as a whole will have to be checked. A thorough flushing will show if there are any build-ups of sludge causing a restricted flow of coolant and a pressure-test of the complete radiator unit could well reveal other faults which cannot be seen.

To flush the radiator, remove the bottom hose and pour water into the filler, preferably using a hose.

Dirty grille

When the fan induces cool air through the radiator, it also includes all the dirt in the atmosphere with a special ration of sticky, oil-like, fractions from the exhausts of diesel-engined vehicles. This sticky content will trap dust and dirt to form a sludge which effectively lowers the cooling area of the radiator to result in over-heating.

Shifting dirt of this sort is difficult if it has built up to any

COOLING SYSTEM

great extent. One effective method is to squirt a degreaser (such as Gunk or Jizer) into the honeycomb from the engine side of the radiator and leave it for quarter of an hour or so to work its way into the dirt. Then use a hose (also from the engine side) to drive the dissolved dirt out.

The top hose

In the cooling system chain, the top hose is always the weak link. The prudent motorist (or one who has suffered from a rash of burst top hoses) will ruthlessly renew it every other year regardless of its condition but even a fairly new hose must still be checked carefully.

The majority of top hoses today are the moulded type but the more flexible convoluted pattern is still popular. In either case, the tightness of the securing clips must always be tested. While it will usually be found that the clips can be taken up a tweak or two, avoid the temptation to use a long screwdriver and a lot of muscle because over-tightening will tend to make the clips cut through the hose and the wire-clip in general use tends to tighten in an oval rather than a true circle which could, in extreme cases, cause a leak to start as the oval shape becomes more pronounced.

Purists prefer the worm-drive clips which retain a truly circular shape no matter how much they are tightened but many moulded hoses have grooves made in them to accommodate the wire clip only.

A sign of hose decay is the tendency for the ends to bell-out very slightly—a tendency encouraged by over-tightening the clips and this should be a warning to renew the hose because wear is starting to take place and wear in a hose seems to increase at a rate of geometric progression.

Cracks and splits in the wall of the hose are also danger signs of the red-alert category and renewal is the only answer. It is possible, always, that the hose will hold together perhaps for months but it will be operating right on the edge of disaster and the cost of a replacement is even today low enough not to justify the risk.

With a convoluted hose which may not show cracks quite so readily as the moulded type, the acid test is to squeeze it so that it is collapsed. The same test can be made to a moulded hose.

The thermostat

The next item in the cooling-check chain is the thermostat—the automatic valve that stops the circulation of water until a pre-determined temperature has been reached.

The bellows type of thermostat will, in the event of failure, remain in the open position. This will have the effect of delaying the build-up of heat in the engine causing a higher-than-normal petrol consumption, increased engine wear and mediocre heater operation.

If a temperature gauge is fitted to the car, a correctly functioning thermostat will be obvious by the gauge staying at the bottom end of the scale for half-a-mile or so and then climbing, fairly rapidly, to the normal position. If the needle starts to indicate more slowly than usual and fails to reach its normal position, despite a long warm-up period, then the thermostat is defective.

Where there is no temperature gauge, the give-away clue is the time it takes for the heater to cut in and a low heat output all the time.

The wax element type of thermostat used in most modern systems, fails in the closed position to cause over-heating fairly quickly and attention is needed right away.

In both cases, a new thermostat is the only answer and, once again, it is as well to note the car's engine number to be sure of obtaining the correct type.

Fitting a new thermostat

The thermostat is found at the engine end of the top hose in a chamber usually dome-shaped (fig. 3). The dome is secured to the main base of the component with screws or nuts which have to be removed (fig. 4). Doing this will involve the loss of some coolant, so remember to drain the system partially in the same way as when changing a hose.

With the dome removed, the thermostat itself can be hooked out of its housing with a pair of pliers. Take a careful note when removing it of the way round it is fitted so that replacement is simplified (fig. 6).

The thermostat dome may stick sometimes and if it will not come loose in the ordinary way, a thin knife blade should be carefully inserted between the dome and main body to break the seal of the gasket. Never use a screwdriver blade for this job because of the danger of distorting

Replacing a faulty thermostat

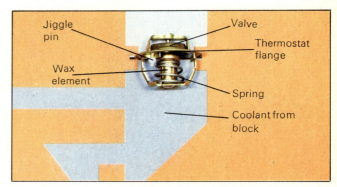

3 A wax element type of thermostat shown here in relation to the waterways. Look for a dome-shaped cover on the engine block

4 To remove a thermostat, first drain the system, remove the top hose, then undo the nuts securing the thermostat cover

5 Carefully prise the thermostat housing cover free from the cylinder block taking care not to damage the joint faces

the dome. Once the tip of the knife has been inserted, work it round the perimeter of the joint a little at a time until the gasket-seal has been sheared.

It will then be necessary to scrape all the bits of gasket off the faces of the dome and housing (figs. 7-8). This is most important because the gasket is usually fairly thin and any bits left could cause a leak.

Just a smear of gasket cement is needed to hold the new gasket in place, the thermostat is slipped into position, the dome replaced and the unit tightened down by working all round the nuts or screws in rotation tightening each a little at a time to avoid any chance of distorting the dome.

Core plugs

A further leak check is on the core plugs. These are fitted, usually, about halfway down the cylinder block with another at the rear of the block. There may also be a core plug in the cylinder head on some engines. The core plugs are there to block the passages left in casting process of engine manufacture and they seal off the cooling system. They also act as a crude form of safety-valve in the event of the engine freezing.

This cannot be relied on, though, and the core plugs should never be regarded as a complete safety factor. In any case, the bother of fitting a replacement far exceeds the cost and relative inconvenience of adding antifreeze to the cooling system.

Even without the stresses of cold weather, a core plug can leak. It is easy enough to spot such a leak through the stains left. It is sometimes possible to use a light hammer to tap the perimeter of the plug very gently to push it fully back into place but this is taking something of a chance and the only certain cure is to fit a replacement.

The worst part of the job is the inaccessibility of the plugs and it may be easier to work from underneath the car. It may also be necessary to remove other components masking the plug. Core plugs are not expensive and it is worth while buying two just in case one is spoiled.

The easiest way of removing the old plug is to use a punch or nail in the middle of the plug and tap is sharply with a hammer to distort it so that it falls out. Then clean the perimeter of the hole with a piece of emery cloth to shift any deposits, smear a little jointing compound round the inside of the hole and press the new plug squarely into place. Tap it all round its perimeter, very lightly, to persuade it evenly into position until it is fully home. Leave it for an hour so that the jointing compound can set and then refill the cooling system.

Reaching the core plug at the rear of the engine is the really tricky job as there is not much space. It may be necessary to lift the engine and some enthusiasts have gone so far as to cut a hole in the bulkhead to reach the plug.

If there is a monumentally difficult plug to treat and the leakage is not too bad, it is worth trying one of the many leak-seal compounds available. These are put into the cooling system and—hopefully—reach the site of the leak to plug it. They do have a fair degree of success but they cannot be guaranteed to work every time. It may be worth trying them in specially difficult circumstances as their cost is not excessively high.

Heater hoses

In the entire cooling system, the heater hoses are taken for granted more than any other item. Yet they are subjected to stresses of the usual sort and sometimes even more so because of their positioning. To make the engine compart-

6 Lift out the thermostat from its housing in the cylinder block, taking note which way up it is

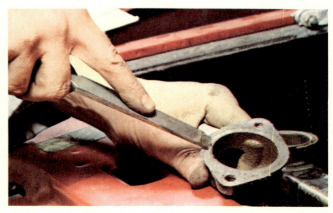

7 Carefully scrape off all the bits of old gasket and cement from the cover joint face taking care not to damage it

8 After plugging the thermostat housing hole with rag carefully scrape the joint facing clean

9 Prepare the new gasket for the thermostat cover by removing the excess material from the pre-punched sheet

COOLING SYSTEM

ment nice and tidy, heater hoses may be led along the cylinder head area where a great deal of heat is generated and oil contamination is possible.

A visual check of the heater hoses will show just how much dirt has accumulated on them and it is difficult to find a better test than simply squeezing them between thumb and finger. In a surprising number of cases, the hose will collapse in a soggy mess whereupon replacement becomes urgent. As a temporary measure, the heater connections can be plugged at the engine but this means there is no warmth in the car.

Heater hoses are usually a mixture of orthodox hose and moulded bends and are usually sold in sets which is not a bad idea because, if one hose is in trouble, it is likely that its opposite number will also be suffering.

Fitting new hoses is usually a fairly simple job because they tend to be of the push-on type with a spring-wire clip holding them in place. In some cases, a worm-drive or wire clip may be used.

The only snag likely to arise is the possibility of an air-lock being set up in the heater system. This can usually be avoided by connecting up the hoses and leaving the end of the hose coming from the hot-water take off point from the engine free. Hold it upright and trickle water into it until it is full and then put your thumb over the end and whip it on to its connection as quickly as possible and with as little water loss as possible.

If the air-lock persists try disconnecting the return hose at the engine end and holding it down below the level of the bottom of the radiator. Run the engine at an idling speed until water, without any air slurps, emerges. Connect the hose and refill the system.

A good way of checking for leaks calls for a completely clean engine area and a hot engine. With the bonnet up and the engine idling, take a look at every connection in the system to spot dribbles of coolant. A helper in the driver's seat to blip the throttle might show up a leak not immediately apparent at low speeds and this routine will also help pin-point a flapping fan belt or wobbling water-pump spindle.

It must be emphasised that the engine must be allowed to cool off before draining and making any repairs because the water will be near boiling point and the system under pressure—a combination which could result in serious scalding if this mistake is made.

Antifreeze

Another important point to remember concerns topping-up in cold weather. Any loss of coolant means a corresponding loss of antifreeze and a consequent lowering of the protection to the cooling system. It is always possible to err on the side of generosity by adding, say, half a pint of antifreeze but a more scientific method is to take advantage of a garage hydrometer which will tell the exact proportion of antifreeze in the system.

The majority of antifreeze solutions on sale today are suitable for cast-iron and light-alloy engines but always check with the label on the container to be on the safe side. In the normal course of events, antifreeze rust inhibitors remain active for a full year, some stay active for longer, but the system should be drained and flushed when the antifreeze has reached the end of its recommended life.

Running repairs

Finally, what can be done if the cooling system breaks down on a run?

A split in a hose can often be parcelled up with plastic insulating tape or other similar material. Strips of cloth, well wrapped round the split and secured with string can be tried as can anything that will parcel the damage sufficiently to prevent water gushing out.

Water pumps can fail at any time and there is nothing that can be done about bodge-repairs. All that can be suggested is to press on carefully and, if possible, keep filling the radiator to replace the coolant lost.

A thermostat jammed in the closed position is a major problem because the water in the top of the engine will boil and it will take an interminable time for the engine to cool off again. It is certainly worth while taking the defective thermostat out and running without it but great care will be necessary to avoid damaging the gasket. Leaving the radiator filler cap off to keep pressure down will help minimise any leak from the thermostat housing and it is then up to the driver to keep alert for any signs of overheating.

Replacing the water lost in almost any form of cooling system fault is always a problem. A small amount will be available from the screen-washer jar otherwise it is a matter of trying to find water from a ditch, a pond or a house along the road.

Unless the fault involves an immediate total loss of water, it is usually possible to keep going by constantly topping-up the radiator but it must be strongly emphasised that driving on a dry engine can lead to the engine becoming totally ruined.

A smashed engine will not be accepted in a reconditioned-engine deal and the cost of a new engine is high. It is up to the individual to balance the factors involved and make a personal decision whether to take a chance or endure discomfort of one degree or another.

10 After lightly smearing both sides of the gasket with gasket cement, put it on the housing joint face in position

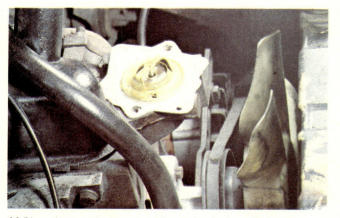

11 Place the new thermostat on its seating in the cylinder block and finally fit and securely tighten both cover and hose

Checking the air cooling system on a Volkswagen 1300

12 On the air-cooled VW, check the tension of the belt. It should yield about 13mm (½in.) under firm pressure

13 If the tension is wrong, first remove the nut from the end of generator shaft, then the outer flange and washers

14 The spacing washers which are between the pulleys are for adjusting the belt tension and any spares are fitted outboard of the pulley flanges

15 Having inserted a screwdriver in the hole in the inner flange to hold the flange in place, tighten the generator shaft nut securely with a spanner

Air-cooled systems

Air being free and readily available at all times, in a form that can be instantly used irrespective of weather conditions, should be the ideal cooling agent. There should also be a total absence of the problems concerned with water cooling such as anti-freeze solutions, leaks, corrosion and the use of sealing compounds on gaskets.

Technical advantages such as more rapid warm up of the engine coupled with the overall simplicity resulting in fewer breakdowns again have been claimed as the advantages of this system.

In all, a quite formidable list, so much so that one would imagine the result would be air-cooled motor cars. However the vast majority of motor cars currently being manufactured are using water-cooled engines. The real snags with air-cooling are the enormous volumes of air needed for satisfactory cooling under all conditions, and this requires a fan running at a high speed and thus consuming considerable engine power. Satisfactory cooling of small in-line four cylinder engines for use in front wheel drive cars is extremely difficult with any form of air cooling because the rate of air flow needed results in unacceptably high power loss. Also the high general noise level is not likely to be accepted today.

The undoubted success of the Volkswagen 'Beetle' in the past with a ducted fan air-cooling system probably was derived from its designer using an opposed piston flat-four cylinder layout.

The servicing for the air-cooling system can be simple but as with all machinery faults can develop.

An engine overheating can be caused by a variety of problems. A very common cause is a slipping fan belt, this naturally will require adjustment since it has to drive both generator and fan.

Adjusting the fan belt (Volkswagen 1300)

The correct tension for the fan belt is when the belt yields about 13 mm (0.5in.) when firm pressure is applied to it between the pulleys (fig. 12). The first stage in belt tensioning is to remove the outer generator flange as shown (fig. 13). A screwdriver is inserted in a hole in the inner flange and supported against the upper generator housing bolt, then unscrew the nut and the assembly will come apart as shown. The packing washers are used to alter the distance between the two pulley flanges and thus tension the belt. Any slackness needs removal of spacers between the flanges, and tightness requires the addition of spacers. Any spare spacing washers which are not actually used between the two flanges must be fitted on the outside under the nut (fig. 14) for safe keeping. Retighten the flange nut (fig. 15) and re-check adjustment. 13 mm yield on all except 1971 models when 6 mm applies—they fitted a stronger belt that year. The belt must be replaced if removal of all spacers between flanges still leaves it slack, and never lever the fan belt over the pulley flanges or both the pulley and the belt will be damaged.

COOLING SYSTEM

Air control flap adjustment
If having checked and adjusted the fan belt, the engine continues to overheat, the most likely cause of the problem is the control flaps failing to open. The air control flaps along with the thermostat are fixed to the crankcase.

It is assumed that the cylinder cover plates and fan housing have been installed with the control flaps in position in the bottom right-hand and left-hand corners of the housing as is usually the case. Firstly screw the thermostat on to its connecting rod from below but do not secure to the bracket. Loosen the nut securing the thermostat bracket (fig. 17). Press the control flaps into the open position and move the bracket until the top of the thermostat is touching the upper part of the bracket (fig. 18). Tighten the bracket nut and check the operation of the flaps by moving the thermostat up and down. Tighten the thermostat securing screw (fig. 21) after checking that the trapped boss fits snugly in the bracket (fig. 19).

Cooling air leaks
Cooling air leaking out through the cover joints and sparking plug covers can cause overheating. Check the sealing caps around the spark plugs.

Caked dirt on the sump, oilcooler and especially on the cylinder cooling fins can cause serious overheating and can only be cured by thoroughly cleaning the system.

Engine tune
Just as faulty carburation or heavily retarded ignition timing can cause engine overheating, a blocked or choked exhaust system can also cause trouble and the only remedy is the replacement of the exhaust system.

16 To get at the thermostat remove the screws securing its cover to the silencer box and the engine crank case

17 With a suitable spanner loosen the nut securing the thermostat bracket to the engine crank case

18 After pressing the control flaps into the open position, move bracket until the thermostat top touches the upper lug

19 After testing the flap operation, check that the boss on the bottom of the thermostat fits snugly in the bracket hole

20 Screw the thermostat securing screw in finger-tight making sure that the lower boss does not slip out of the bracket

21 Fully tighten the thermostat securing screw with a spanner and then refix the thermostat cover plate

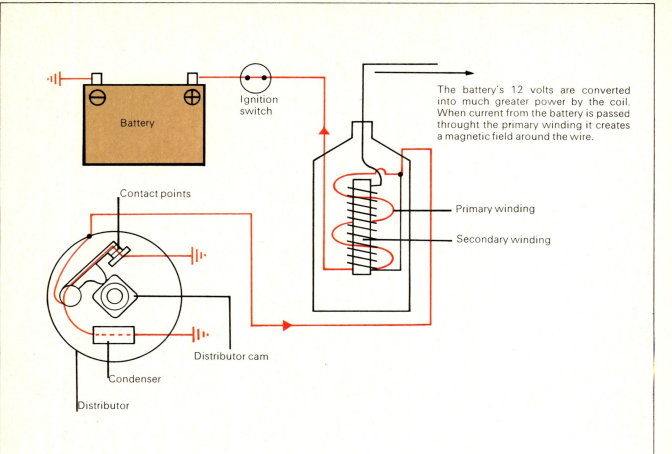

The battery's 12 volts are converted into much greater power by the coil. When current from the battery is passed throught the primary winding it creates a magnetic field around the wire.

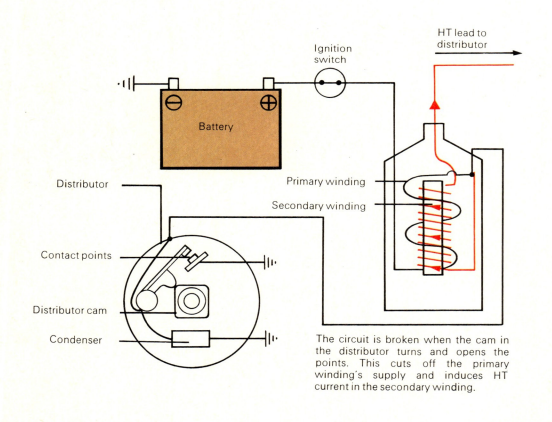

The circuit is broken when the cam in the distributor turns and opens the points. This cuts off the primary winding's supply and induces HT current in the secondary winding.

CIRCUITS

Checking the electrical systems

The modern car has quite a complex array of electrically-operated components and accessories, most of which operate for many hours without any trouble. Nevertheless, just as a car's mechanical components tend to deteriorate over a period and require regular inspection and servicing, so the electrical systems may gradually develop faults and accordingly should be checked at the same time.

2 Two common car fuses: (right) a 'standard' fuse and (left) a 'continental' fuse. Both types are normally continuously rated

Service Checklist

13 Check cooling components for leaks and blockages. Check condition and tightness of fanbelt.

★ 14 Check operation and condition of all electrical equipment. Adjust headlamp settings if required.

15 Check all steering linkages for excess wear and movement; repair/replace as necessary. Examine driveshaft boots and steering rack gaiters for security and splits (where applicable).

A convenient and sensible time to make sure that all the electrical items on your car are working correctly is the 12,000 mile service. In most cases these tests are very simple: a component either works or it does not. If the latter is the case, then the procedure for finding out exactly where the fault lies is quite straightforward and well within the scope of even the most inexperienced d-i-y mechanic.

How car electrics work

Before you can effectively carry out any practical checks on the electrics of your car, it is important that you understand the principles involved.

In order to operate any electrical item, current must be conducted to it from the source of supply and then back again to form a complete circuit. At the source (the battery or generator on a car) one conductor must be connected to the positive terminal and the other to the negative terminal. The unit being supplied must also have two terminals—in some cases they are interchangeable while in others they are identified as positive and negative and must be connected accordingly.

In many wiring systems in other fields, for example the normal domestic one, two separate wires are used: one for the 'outward' run of current to a unit and another for the 'homeward' run back to the opposite terminal of the source. A similar system was used in the earliest cars but most modern cars use what is known as the 'single wire' or 'earth return' system. (One notable exception to this rule is a glass-fibre-bodied car where a separate earth lead runs either to the car's chassis or the negative terminal of the battery.)

With an earth return system, a carefully insulated wire is used for the first, live half of the circuit which carries the current to a unit. The metal of the car body takes the place of the second, earth wire and thus completes the circuit to the power source. Short lengths of cable are used where necessary to bridge any gaps in the earth side—for example, to connect the engine (which is usually insulated on rubber mountings) to the main frame.

Two factors are important in any car electrical system. First, there must be excellent electrical contact between the various parts of the vehicle used as conductors. And second, there must be perfect insulation of the live (wire) side of the circuits from any part of the earthed frame. The only points on the live side that may be exposed are the terminals at the end of the live wire. Even then, these joints should be protected from accidental contact as far as possible – either by the use of an insulation sleeve or a piece of insulating tape.

Wiring faults

These two important factors—contact and insulation—produce two common types of wiring fault. Failure to get good contact between conductors obviously prevents or hinders the operation of a circuit because it prevents some or all of the current flowing through it. Corrosion, dirt or moisture can all cause problems of this sort, as can a simple break in the wiring which results in an 'open circuit'.

Failure in the insulation of a live wire, on the other hand, allows the current to leak directly back to the return circuit. This is called a 'short circuit' and is dangerous for two reasons. First, the current will not reach the intended component (which therefore cannot function) and second, since it has no work to do, more current than can safely be carried flows through the circuit as far as the short. At best this results in a blown fuse and at worst in a fire as the wires overheat.

Car wiring practice

Supply wires to various components on the car are generally bound together where they share a common route. The resulting group of separate cables is known as a 'harness' or 'loom'. Fortunately, manufacturers make provision for

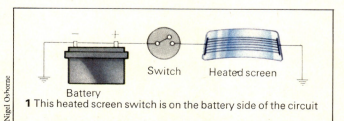

1 This heated screen switch is on the battery side of the circuit

tracing the individual circuits either by colouring or numbering the insulation of each cable. The location and destination of each wire is included in a wiring diagram which is often to be found in your owner's handbook.

Replacing a wire

If you suffer a burnt-out wire, a short or open circuit in a lead or simply come across a section of wire which looks very shabby and liable to fail, you will need to replace it. Often this entails renewing a section which runs in either the main or a subsidiary loom.

Whether you are renewing or simply investigating a particular wire in a loom, you will need to open up the part of the loom the suspect wire runs through. Probably you will have to disconnect all the wires from their connections at one end (make sure you label them for easy reconnection). Then pull the loom out from its fittings on the car bodyshell.

When the loom is in a convenient position to work on, carefully slit it open with a sharp blade (fig. 7) taking care not to damage the cables inside. You can now reach the faulty wire; if it has melted, check carefully that it has not damaged any of the other wires. Always replace damaged leads with new ones of the same thickness (see below). It is also a good idea to replace a burnt-out yellow wire, for example, with a new yellow wire as this will keep the colour coding of your car intact.

When changing a wire, do not pull out all of the wire in one go; instead work from one end, a few centimetres at a time, pulling the old lead out and pressing the new one in (fig. 8). This will preserve the shape of the loom. Re-band each section, using wide insulating tape (fig. 9). It may be possible to transfer the connectors on to the new lead if they are the crimped or soldered type.

Wiring grades

Most ordinary car electrical circuits use 32/0.2 mm wire; heavier work such as feeding headlamps is usually handled by 28/0.3 mm wire while 44/0.3 mm wire is invariably used for DC charging systems. The figures describe the number of strands in the cable and the thickness of each one. Older British cars will have Imperial-rated wiring, the equivalent grades to the above sizes being 14/0.012in., 28/0.012in. and 44/0.012in. respectively.

Fuses

Fuses are the deliberate weak link in an electrical circuit and their principal job is to protect the wiring and the electrical components. For example, if a short occurs, the fuse wire is designed to melt so breaking the circuit before anything else is damaged. If a fuse is blown, it will have blown for a good reason so always try to find the fault before renewing the fuse. Do not merely change the fuse and hope for the best; and never, except as a 'get you home' measure if you are sure that nothing is wrong with the circuit, substitute wire, silver paper or anything else for a proper fuse of the correct value.

Value of fuses is where confusion can sometimes arise. Some fuses (fig. 2) are what is known as 'continuously rated'—that is, marked with the current they are designed to carry continuously, but they will blow at double this figure. On the other hand, other fuses are 'blow-out' rated which means they will carry continuously only half the current they are marked with. To make matters simpler, some fuses are marked with both the 'continuous' and the 'blow-out' rating, but if the fuses in your car are not, always play safe by buying replacement fuses at your local franchise dealer.

If you are not sure what value fuse to fit in a particular circuit, first look in your owner's handbook. If the handbook is no help, then you can work out the nominal 'continuous' rating of a fuse by dividing the wattage of the relevant component by the voltage of the battery (normally 12 volts). For example, to find the 'continuous' rated fuse for a 60w headlamp bulb, divide 60 by 12—which means you need a 5 amp fuse.

Electrical testing equipment

Before you start work you will need both a test lamp (12 volt or 6 volt to match your car's system), which lights up if connected into a live circuit (fig. 6) and also a circuit tester (fig. 3), which consists of a bulb with its own battery. These items are used for different jobs and are not interchangeable. For example, if you pass a full 12 volts through the circuit tester it will probably blow its bulb. Both should ideally be equipped with a probe at one end and a crocodile clip at the other. In addition, a couple of loose lengths of insulated wire will be handy for bridging circuits and so on.

Many of the tests described below require you to connect one end of your circuit tester to the car's earth. Finding a good earth point on a car, other than the battery's earth terminal, can be tricky. A layer of paint acts as an insulator, so you may have to scratch through the paint to bare metal in some place where it will not matter, such as a door pillar. An unpainted bolt or screw head is probably a fairly good earth. It is worthwhile filing through to bright metal on any earth point you choose, to make sure that dirt or oil are not preventing current from flowing.

Remember when testing any electrical circuit that there is a considerable amount of energy in a car battery. If you accidentally let a loose live wire touch the chassis there will be a spark and maybe a bang as the wire shorts out. If your hand is nearby it may get burnt, so take care.

3 To check that your circuit tester is working properly, touch its lead and probe together. The bulb should now light up

4 To find out whether a fuse has blown or not, remove it from its holder and use your circuit tester to check its continuity

CIRCUITS

How to trace an electrical fault

One of the fundamental requirements of successful electrical checking is to be careful and methodical. Try to stick to a logical sequence at all times and, if you find any faults, you should not have too much difficulty in isolating the cause of the trouble.

If a component does not work, first check that the fuse carrying the current supply to it is intact. Remember that on many cars, groups of components are supplied through a common fuse. In fact, early Leyland Minis have only two fuses to serve the whole car—one fed through the ignition switch and the other direct from the battery. Therefore when one electrical component malfunctions it is easy to tell if the fuse is to blame simply by switching on one of the other units supplied by the same fuse.

On some cars the fusebox itself is marked to show which components are supplied by each fuse. On others, you should find the information in your owner's handbook. It is worth making a note of this to save time in case something goes wrong at a later date.

If the fuse appears intact, check also that neither the fuse holder nor the fuse has corroded and that the relevant fuse is firmly in position. If you suspect the fuse to be blown, either substitute it with another fuse of the same rating or remove it and test its continuity. To do this, clip the lead of your circuit tester on to one end of the fuse and touch the probe to the other (fig. 4). If the lamp does not light the fuse is blown. In that case, look for the short circuit which may have caused the fuse to blow.

Possible causes of a 'short' are a fault in the component being supplied, corrosion around its terminals, a lead which has come adrift and is touching the chassis, or a wire whose insulation has chafed against the chassis. The only way to isolate the fault is to check the circuit through, looking for faults, or to test the component separately.

If you discover the fuse to be undamaged, your next step is to trace the path of the supply wire from the switch right the way along to the component itself. Make sure that all block connectors, junctions and unions are securely in place. This may seem obvious but a surprising number of electrical 'faults' can be cured in this way.

Should the component still not work at this stage, the fault could lie in the component itself. This is where your spare leads come in; connect the component's live battery terminal (positive on a negatively-earthed car) direct to the positive battery terminal and join the body of the unit (or its earth terminal if there is one) to a good clean earth point—or the negative battery terminal if it is conveniently placed. If it does not work when connected like this, the component itself is faulty.

The test lamp can now be used to find out where a circuit is breaking down if you have established that the fuse and the actual component are both good. Switch on the relevant circuit and connect one lead of the test lamp to earth (fig. 5). Commencing at the supply terminal of the non-functioning unit, work back along the supply lead with the probe of the test lamp. (If the probe is sharp enough, you can penetrate the insulation of the lead without damaging it—otherwise you can test only at cable junctions and terminals.) As soon as the lamp lights, stop; the faulty section of wiring will be that you have just tested (fig. 6).

If the lamp lights anywhere on the component side of the fuse, the fuse must be in order. Similarly if it lights right on the supply terminal of the inoperative component either the component itself or its earth connection must be causing the problem.

How a car lighting system works

A high proportion of the electrical components on any car are lights or lamps of one kind or another. There are about 18 bulbs on the average car—and many more on those models which have several warning lights on the dashboard.

Most lighting circuits are similar and all of them incorporate a switch which lets the power through when you operate it—or, in the case of warning lamps, when something like a drop in oil pressure operates it. The wire from the switch carries power into the lamp holder where it feeds one terminal of a bulb. The current flows through the filament of the bulb, heating it and causing it to glow, then returns to earth to complete the circuit. It may reach earth via another terminal or through the brass casing forming the base of the bulb. In either instance, the return path then normally goes via the body of the lamp assembly through its mounting screws or bolts to earth.

So, switch on all the lights one at a time and see if they work. Back the car close to a wall or another car and look for the reflection to check your stoplights.

SPECIFIC CHECKS

Sidelamps

If a sidelamp bulb is not working, then first have a close look at the bulb itself. You may be able to see whether or not the filament has broken. If the filament is not intact then the bulb has 'blown' and must be replaced. To test a bulb, either substitute another which you know to be good or hold its casing to earth and join its contact direct to the battery live terminal (figs. 10 and 11).

If the bulb works now, but not when you put it back in its holder, use your circuit tester to check the continuity of the connection between the case of the bulb and its holder, as corrosion or dirt can sometimes upset this. Simply clip

5 Use your test lamp to check the wiring to a non-functioning component. Clip one lead of the test lamp to a suitable earth

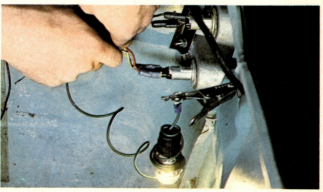

6 Switch on the circuit and work back along the supply lead with the test lamp. The bulb will light up if the wiring is in order

the tester to a metal part of the holder (fig. 13) then touch the probe to the bulb's brass side, with the bulb in place, taking care not to allow the probe to touch the bulb holder (fig. 14). If continuity is good the tester will light. A similar procedure can be used to check the earthing of the holder; this time connect the tester lead to earth and touch the probe to the bulb holder again (fig. 15).

Supply to the bulb holder can again be checked with the test lamp. Connect one end to a good earth point, such as bare metal on the chassis, and use the probe to touch the contact in the holder. With the circuit switched on, the test lamp should light. If it does not, follow the procedure described earlier to check the supply all the way back to the fuse, if any, or to its source.

A lighting switch, or any switch for that matter, can be checked very simply. All you have to do is bypass it—pull the connectors off its terminals (after noting where they go) and touch them together. If this makes the circuit work then the switch is clearly faulty.

Headlamps

Headlamps are a little different from other lights but the principles of checking are the same. If both (main and dip) filaments of one lamp fail to light, the fault is likely to be in the sealed beam unit. If either both dipped beams or both main beams fail to operate while the other pair of beams are working correctly the trouble almost certainly lies at or within the dipswitch. If both the headlamps are totally inoperative, a problem at the main lighting switch (usually on the dashboard) or a blown fuse are the strongest possibilities. Check also that you have not got a flat battery.

Dimness of lights is often caused by bad earthing and, even if not shown up by the circuit tester check described on page 107 and above, it should be investigated closely. Apart from a possible fault in the supply lead, remember that an old bulb will suffer from blackening of the inside of its glass envelope and therefore its output intensity will be reduced as time goes by. Quartz-halogen bulbs are less affected by this.

Direction indicators

The circuit for these is similar to that of other lamps except that they receive their current via a thermal make-or-break switch contained in a canister—the flasher unit. Check that your indicators all work and that the rate of flashing falls within the legal requirements: between 60 and 120 flashes per minute.

Slow flashing on one side only is invariably caused by an earthing problem somewhere on that side. Extra fast flashing on one lamp (with or without the dashboard warning light) indicates that the other bulb on that side (front or rear) has blown. Too fast flashing all round is almost certainly due to the flasher itself.

If the indicators function correctly, but their warning lights do not operate, check the warning bulbs or their bulb holder connections for a poor earth. If the lights still do not work, the flasher unit is faulty and should be replaced.

Total failure of all indicators may be due to a breakdown of the flasher unit or it may be a blown fuse or a break in the supply circuit. If the fuse has not blown check the flasher unit as follows. Pull the input and output wires off the canister terminals (marked either B and L for British cars or + and Com for others) after noting which wire goes where. Next, link the two wires together—a paperclip pressed into their connectors will do—and operate the indicator switch first one way and then the other. If all the bulbs on one side light when the switch is thrown to that side, it is the flasher unit which is at fault.

Indicators should always be in working order for reasons of both safety and legislation.

Stop lights

Stop lights, like indicators, often suffer from earthing problems. Otherwise, brake light switches themselves can occasionally give trouble—you can find out whether it is the switch at fault by following the procedure already described; that is, pull the wires from the switch and join them together (figs. 16-17). If the brake lights now work with the ignition switched on, then the brake light switch will need replacing.

Some brake light switches operate via hydraulic pressure in the brake lines and cannot be adjusted (adjusting the brake light switch alters the pressure needed on the brake pedal to operate the brakelights). The other type of switch is a mechanical one and is usually operated by the brake pedal depressing a plunger when pressed itself. This type

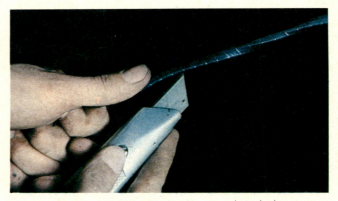

7 If you find that one of the wires in a loom needs replacing, first carefully slit open the loom with a sharp knife or blade

8 Working a few centimetres at a time, pull out the old lead from the loom and press in a new lead of the same rating and colour

9 When the new lead is in place, re-band the loom with wide PVC insulation tape and return the loom to its original location

CIRCUITS

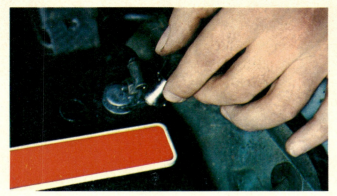

10 To test if the filament of a bulb is intact, first rest one of the bulb's bayonet pins against the battery negative terminal

11 Now use a length of insulated wire to join the bulb's contact and the battery live terminal. The bulb should now light up

12 Checking the connection between the case of the bulb and its holder is easy. First pull back the rubber protecting cover

13 Then pull the bulb holder out from the lamp assembly and clip the circuit tester lead on to a metal part of the bulb holder

14 Without letting it touch the bulb holder, touch the probe of the circuit tester to the bulb's casing. The tester should light

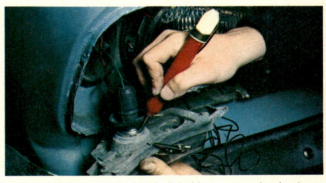

15 To check the earthing of the bulb holder, connect the circuit tester lead to earth and touch the probe to the bulb holder

16 The operation of any switch can be checked by bypassing it. To check a stop light switch, first pull the wires off the back

17 Then use a paperclip to link the two wires together. If the brake lights now work, with the ignition on, the switch is faulty

can be adjusted if it is not working properly either by screwing the switch in or out of its mounting or by simply bending the mounting bracket. Mechanical stop light switches have a locknut which have to be undone before adjustment; do not forget to re-tighten it afterwards.

Other lights

Interior lights may be checked in the same way as the other types of lights except that in order to operate automatically from switches in the door pillars, the switches have a continuous live supply, rather than only working when the

18 Courtesy light plunger switches are usually a press fit into the door pillar but can be held in place by cross-headed screws

19 These plunger switches are prone to corrosion. Press in the plunger and run a piece of abrasive paper between the contacts

ignition is switched on. Bonnet, boot and glovebox lights may work in the same way. The small plunger switches (fig. 18) are prone to corrosion but can usually be removed and cleaned up without difficulty. In some cases the switch is simply a press fit into the door pillar, and can be removed by prising it out with a screwdriver. Make sure not to lose the wire down inside the pillar if you disconnect it.

Instrument lamps may be connected to or separate from the main lighting circuits. Exactly the same procedures apply to checking them as to the other lights.

Rear screen heater
The circuit for a rear windscreen heater is straightforward and, in the event of failure, can be checked as for a lighting circuit (it is a good idea to have the heated screen ignition-controlled as it needs a lot of current and will run a battery flat very quickly if left on unintentionally).

The warning lamp may burn out without affecting the operation of the heater element so if the bulb goes out, look at this possibility first. Use your test lamp to see if current is reaching the live element terminal (with the probe on the terminal, the other end to earth, and the circuit on, the lamp should light). Conversely, the warning lamp may continue to glow when the element is not receiving its full current. Again, check with your test lamp and look for poor connections in the supply lead or, more likely, a bad earth.

Wipers and washers
Apart from mechanical faults, such as rack or gear wear, wipers are generally reliable. Total failure is often a result of a blown fuse, though it is possible for a motor to burn out through overheating, especially if the wiper blades have been continuously run on a dry windscreen. Use your test lamp to check if current is reaching the motor if you find the fuse to be good yet the wipers will not run.

Electric screen washers simply use a small electric motor to drive an impeller. This can overheat if used for too long (particularly when out of water). Again, check the supply to the motor and its earth.

Horns
A horn that does not work or emits a peculiar, strangled noise has usually nothing more wrong with it than lack of use. Therefore, have a friend press its switch while you give the horn itself a series of sharp blows. This will unstick the 'gummed-up' contact breakers inside the horn and so cure the fault.

Many horns are earthed through their mountings with the switch on the battery side of the circuit. Some others, though, have a continuous live supply (through a fuse) with a separate wire leading to earth via the switch. On the former type, the mountings are sometimes a problem due to corrosion. Clean them up and make sure that the nuts and bolts are tight. You can check the earth with your tester.

Horn-pushes are subject to a certain amount of arcing—electricity jumping the gap between the contacts, causing sparks and burning—because they draw a heavy current (an intermittent note signals trouble here). Dismantling and cleaning the horn-push is the best cure for this kind of burning but you can often get an old-style horn-push working again by simply twisting the button in its socket.

A horn that is both earthing properly and receiving a good supply of current yet still will not work after you have hit the casing fairly hard may respond to contact adjustment. This is often possible via a screw or nut on the back. If there is no outside screw or if the contacts are badly corroded, you will now need to open up the casing. Sometimes there are retaining screws but more often there are rivets or pressed-over flanges. At this stage, however, it is often simpler to fit a replacement unit; one from a breaker's yard will cost comparatively little. In many cases you may not even need to use a horn from the same model of car, as long as the voltage is the same.

The heater
The heater fan is an electric motor which is fed through a fuse. It may have a simple on-off switch, or a rheostat—a variable resistor—may provide variable speeds. If by switching on the motor nothing happens then use your test lamp as before, not forgetting the fuse. A rheostat can be bypassed in the same way as any other switch to check whether it is faulty. Failure of the heater motor itself is rare, and it is more likely that a connector has come adrift.

Alternatively, your heater switch may have two settings, for high and low speeds. In this case the higher speed is the normal fan rate, while the lower speed is obtained by simply passing the current through a resistor. If the high speed works, but not the lower, suspect this resistor and check it with your test lamp.

Headlamp adjustment
Another vital aspect of checking your car's electrical system is the headlamp adjustment. The most important thing to remember when adjusting your car's headlamps yourself is that you cannot carry out this job accurately enough without the special beam-setting equipment possessed by most garages. However, you can check the adjustment roughly and even an approximate setting is better than nothing.

Remember also that incorrect headlamp adjustment is

CIRCUITS

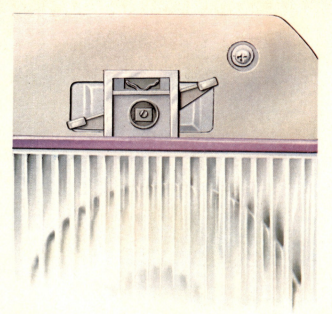

20 The Peugeot 104 uses this design of headlamp load adjuster

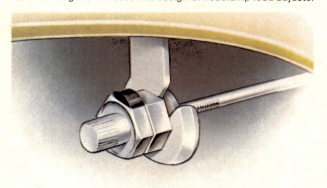

21 This type of load adjuster is fitted to the Peugeot 504

23 Another type of load adjuster. This one is a Renault design

both illegal and discourteous to other road users. Asymmetric (found on most Continental cars) and four-lamp systems are particularly tricky to deal with since both require setting to be made on dipped beam. Always regard your own setting as strictly temporary—to be followed up by a professional garage setting.

Every car manufacturer specifies a slightly different procedure for setting lights, and no two agree absolutely. There is, however, a compromise method which suits cars with round headlamps like, for example, the Leyland Mini.

First find a level area of ground with a vertical wall at least 1 m high. Check the tyre pressures of your car and have one passenger (or equivalent weight) in the rear seat. If you are soon to go on holiday, load the car accordingly. Now drive the car at right angles as close as possible to the wall and facing it. Mark a cross with a piece of chalk on the wall exactly in line with each headlamp centre. Join the crosses with a horizontal line. Next, reverse the car between 8 m and 10 m away from the wall in a straight line and

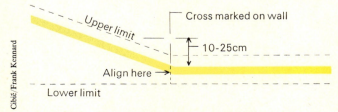

22 Diagram for the adjustment of asymmetric headlamps. The dotted lines represent the upper and lower limits of the dipped beam

switch on the main beams. The brightest areas should be centred a few centimetres below each cross. Cover each lamp in turn to check that the other is right.

If adjustment is necessary, work on one lamp at a time. On some older cars you may have to remove some of the brightwork or the radiator grille to gain access to the adjustment screws. On modern designs, they are either easily accessible from the front or else have been replaced by knurled finger-nuts at the rear, which are reached simply by opening the bonnet.

There may be two or three adjusters, but basically the topmost one is used to govern vertical aim and the side ones give lateral adjustment. Turn these adjusters until the beams point to the desired spots on the wall.

Asymmetric headlamps

Asymmetric beams, which are identifiable by the long finger of light which should be directed along the kerb on dipped beam, must be set on dip. A typical instruction is that having drawn the crosses on the wall as before, drive the car back 10 m. The horizontal cut-off—that is, the top edge of the lowest part of the illuminated area (fig. 22)—should be between 10 and 25 cm below the crosses on the wall. Lateral aim should be such that the point where the extended kerb-side beam begins to slope upwards away from the cut-off horizontal section is vertically in line with the crosses.

Most French cars and some others have a provision for simply adjusting the headlamp beam height (without disturbing the adjustment) to compensate for varying loads. If this takes the form of separate levers (figs. 20, 21 and 23) either in front of or behind each lamp unit, set the beams with a normal load in the car (in other words, with one passenger in the back) and the levers in the light-load position (their more upward-pointing position). If, as on some Citroens and Renaults, there is a continuously variable knob accessible from the driver's seat controlling both lamps, set the beams with no load on board and the lamps again in the highest-aimed position.

Checking the steering

If your steering fails when you are driving at speed, the experience can be at the least unnerving—and at the worst fatal. It pays to make regular checks on the steering mechanism of your car, to reduce the possibility of a sudden breakage.

A thorough check of the steering should be made at every major service. Once slight play has developed in a component, wear can rapidly result. However, as the steering gets worse gradually, you may become used to its condition without realizing how much it has deteriorated.

This chapter summarizes all the checks you should make to see that your steering stays in first-class order. The 12,000 mile service is the ideal time to guard against potential faults.

Start checking by gripping the steering wheel and rocking it up and down and from side to side. You can always turn the wheel to a certain extent, of course, on account of the flexibility of the tyres against the ground, but apart from this there should be no visible free play, and the column should not be loose in its bearings.

Service Checklist

14 Check operation and condition of all electrical equipment. Adjust headlamp settings if required.

★ 15 Check all steering linkages for excess wear and movement; repair/replace as necessary. Examine driveshaft boots and steering rack gaiters for security and splits (where applicable).

16 Examine tyre wear and change wheels round if necessary. Check wheel alignment and tyre pressures.

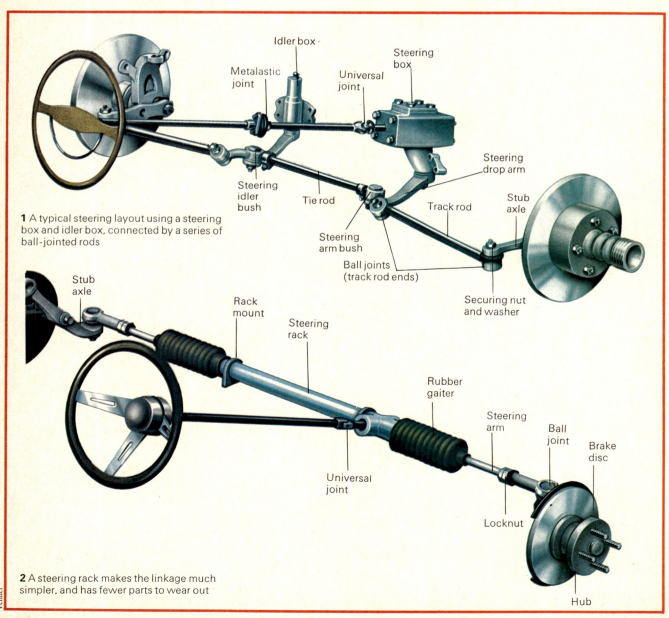

1 A typical steering layout using a steering box and idler box, connected by a series of ball-jointed rods

2 A steering rack makes the linkage much simpler, and has fewer parts to wear out

STEERING

3 Start checking the steering by gripping the steering wheel firmly and rocking it up and down in its bearings

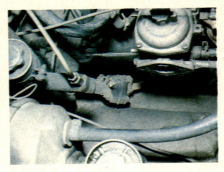
4 Some cars have flexible steering column joints. They can be small universal joints, or Metalastik rubber couplings as shown

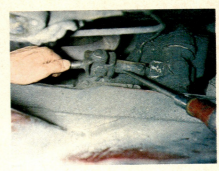
5 Holding one half of the joint steady with a screwdriver, twist the other half by hand to check for any looseness or free play

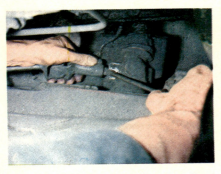
6 Holding a screwdriver against the splined steering shaft, test for worn splines by twisting the joint from side to side

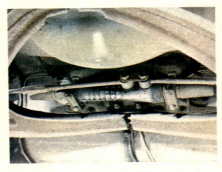
7 The steering rack on this Hillman Imp is shown with the steering on left-hand lock. The elongated gaiter is clearly leaking

8 One method of checking steering rack wear is to grip each track rod in turn, feeling and listening as the steering is turned

9 To check end float in the steering box arm insert a strong lever between the arm and the tie rod, and lever gently downwards

10 The same method can be used to check for wear in the idler box. Look for any signs of deterioration in the joint bushes

11 Hold the tie rod and one track rod at a time, and have an assistant gently turn the steering wheel on full lock either way

Using a tyre tread as an indicator
Next, examine carefully the state of the front tyre treads. These can be a good indicator of the state of the steering mechanism. Turn the steering to full lock in one direction: this will make examination easier.

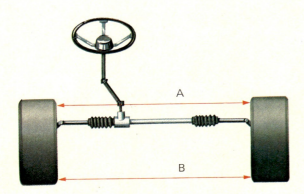
12 If dimensions A and B are similar, the wheels run parallel. If A exceeds B, there is toe-in. If B exceeds A there is toe-out

All cars have a specific alignment figure which determines whether the front wheels run parallel to one another, or whether the leading edges face inwards slightly (toe in), or outwards (toe out). This measurement is called the 'track'. If the car has excessive toe out, the inside edges of one, or both, of the tyres will wear more than the rest of the tread. If the treads are worn on the outer edge, there is too much toe in.

To use the tread wear patterns as an accurate guide to the condition of the steering joints, it is essential that the tyres have always been at the correct pressures throughout the car's life.

When the tyre is worn on the inside edge of the tread, and the wear extends across the tread giving the tyre a tapered appearance, the cause is usually worn king pins or swivel joints. In the latter case, the worn joint allows the wheel to lean outwards sufficiently to cause wear during cornering.

Another danger sign to look out for is wear on either side of one, or both, front tyres. This could be due to incorrect tyre pressures, but if the pressures are known to have been regularly maintained, the most likely cause is a worn steering ball joint.

113

Checking underneath the car

To check the steering linkage thoroughly, the front of the car at the very least must be raised from the ground. Large axle stands are ideal for this job, but if you are fortunate enough to have access to a pit or drive-on hydraulic lift, so much the better. Simply turning the steering wheel from side to side with the wheels on the ground will serve no purpose; the front wheels must be raised.

With the car on level ground, raise it as suggested, having made sure that the rear wheels are securely chocked. Have an assistant turn the steering wheel gently in either direction. It is not necessary to use violent movements: in fact these are more likely to mask any steering wear. Slow movements will reveal any slack in a joint.

Check the steering column flexible joints first. These will either be of the Metalastik rubber variety (fig. 4), or small universal joints (fig. 5). There should be no visible wear in either the joint or the splines connecting the column to the steering box or rack. Smear universal joints with a little grease to prevent dirt and moisture getting in, but do not lubricate rubber joints.

On the majority of cars, the steering gear is visible from below, but on certain models, such as the Wolseley 1500, Riley 1.5 and the Hillman Super Minx, the linkage runs across the engine bay, and is viewed from above. Your car will be fitted either with a steering box or with rack and pinion steering. The steering box has a main link rod running from the steering arm to an idler arm—the box and the relay arm are usually bolted to the chassis rails or subframe. The track rods, with a joint at either end, are connected to the steering or idler arms, and join these to the link on the stub axles, giving a total of six joints.

Steering racks have two ball joints, one on either end, which again are connected to the stub axles.

Carry out a visual inspection of all joints which have a rubber dust boot. If any are damaged, the whole joint will almost certainly have to be replaced, as dirt and grit will probably have begun to destroy the inner cup. It is difficult to obtain rubber dust covers separately, especially as most parts manufacturers now operate a unit-replacement policy. Some older, more expensive cars, such as Bristols, have adjustable ball joints. On these it is possible to add or extract shims to take up wear.

Many cars have grease nipples on the ball joints, but the modern trend is towards sealed-for-life joints. It is not possible to lubricate these. When they become worn, they must be replaced. Check also that the steering box has no oil leaks around the seams and the arm pivot. If the rubber gaiters on the end of the steering rack are damaged, they must be replaced.

With your assistant rocking the steering, watch the rack or box for any sign of looseness on its mountings, and also inspect the idler arm mountings. Look for any slack that may be present in the arm shaft, or on the rack. There should not be any side play between the inner moving part and the housing as it strains against the wheels—this wear can often be detected as a rattle while driving.

One of the most effective ways of detecting small amounts of wear in the joints as the steering is rocked is to clasp your hand over the joint so that both sides can be felt. Lost movement can easily be detected in this manner. Check all joints, including the inner rack joints under the gaiters. You will probably have to turn the steering to full lock for this, so be careful not to trap your hand.

Watch for undue play in all pivot points and suspension joints as the steering is moved; small amounts of wear in the king pins or swivels will probably go undetected, but will show up when the wheels are jacked off the ground.

Leave the car securely supported on its axle stands, and make sure the suspension is on full droop. Check for any

13 Check the front wheel bearings for play by grasping the road wheel top and bottom and rocking it inwards and outwards

14 Hold one road wheel and pull it until it reaches full lock, then swivel it to the other lock. Listen carefully for knocking

15 Supporting the car on the suspension, use a wooden block as a fulcrum and lever upwards. This shows up swivel joint wear

16 The lower swivel joint is situated at the bottom of the kingpin, and is held in place by a tapered pin and locknut

17 The upper swivel joint. Wear in either the upper or lower joints can cause sloppy steering, and may be hard to detect

18 Some ball joints, such as this example on a Jaguar, have grease nipples. These must be protected if the joint is removed

STEERING

looseness in the wheel bearings by grasping each front wheel at top and bottom and trying to rock it. The adjustment and replacement of all types of wheel bearings is largely a matter of varying techniques but, as a guide, there should be virtually no play, whether your car has tapered roller bearings or whether it has ball bearings. The hub should not, however, be so tight that the bearings bind as it rotates. If you are in doubt about how much play is acceptable, check with your local dealer.

A quick extra check on the steering linkage can be done by rocking the road wheel in the steering plane. Grab the tyre and briskly rock the wheel to and fro. The point at which you change direction will show up any obvious looseness by lost movement, and sometimes a clunk or a rattle.

The most important check is to grasp the road wheel firmly at the top and bottom, and energetically rock it in the vertical plane. The wheel often has to be moved with some force to show up wear in the king pins or swivels in addition to any slight movement in the bearings. Even the smallest movement here should be thoroughly investigated. Get an assistant to rock the steering wheel for you so that play can be carefully observed and the exact points of wear located. It is possible for the play to be in the inner and outer suspension pivots as well as the steering, so a successful repair depends on your spotting all the faults in the first place.

For the next check, you will need an assistant to apply the brakes, locking the wheels, so that you can rock the wheels in the rotating plane. This provides a large lever—in the form of the wheel itself—with which to check for worn joints.

As a further check, support the weight of the car on the spring pans, where coil springs are fitted, or the lower suspension links so that the front wheels are about three inches from the ground. Place a long metal rod under each tyre in turn and lever upwards. Be careful not to upset the attitude of the car. This test is useful for identifying any wear in the upper and lower swivel joints (figs. 17 and 16).

On some cars it is very difficult to detect wear in some joints. Certain types of kingpins, such as those fitted to the Morris Minor and Triumph Herald, pivot on a screw-threaded principle. The best way to check these is to dismantle the unit and examine the threads for wear. These joints should always be kept well greased—they should be lubricated every 5,000 km (3,000 miles) to be safe. If such joints snap, the wheel will tilt at an angle, possibly causing loss of control.

Replacing ball joints
Having checked the steering linkages thoroughly, you may well find that a ball joint shows signs of wear, and needs replacing. This is a fundamental job that is easily tackled, but it can be one of the most frustrating jobs for the uninitiated. The joints are attached by a taper pin passing through an eye in the steering arms/links, and secured by a nut and washer or split pin (fig. 19).

Raise the front of the car on ramps. Change only one joint at a time, as it is important that the steering cannot be accidentally disturbed while the part is being changed.

Slacken the locknut on the tie rod (fig. 20).

The time-honoured method of removing ball joints is to undo the securing nut by two full turns, then strike the ball joint to disturb the taper. There is nothing wrong in this method, but it can be an unnerving experience constantly hammering at a joint that apparently refuses to loosen. It can be useful to place another hammer behind the joint, to take out some of the natural flex in the metal. Very often, lack of under-car space, or the close proximity of delicate engine parts, such as alloy sumps, prevents the use of a

19 A typical ball joint, on the outer end of a track rod. This joint is held in place by a castellated nut and a split pin

20 On this particular type of joint, the locknut can be fully slackened without any alteration of the tracking alignment

21 Remove the split pin in the castellated nut. Do not re-use the same pin when you reassemble. Slacken the castellated nut

22 Using a heavy hammer to absorb some of the natural flex of the metal in the joint, strike the joint with another hammer

23 Special separators can be useful if the joints are hard to split. Tap the claw between the joint and the steering arm

24 Carefully tighten the bolt until the tool impinges on the pin, then continue tightening until the joint springs apart

hammer, or restricts the amount of force that can be exerted. In this case, it is better to purchase a ball joint separator. These are available in several designs (fig. 23). Some operate on the principle of scientifically exerted leverage, while others act simply as a forked drift. These tools are well worth consideration, for they are inexpensive and can be used time and again.

The securing nut should be loosened until any nylon inset is clear of the taper pin thread. Tap the claw end of the separator between the joint and the steering arm, and tighten the bolt until it impinges on the centre of the pin. Tighten the bolt gently, but firmly, until the joint springs apart. Take care not to lie immediately underneath the separator, as it may suddenly drop away from the joint as the connection is broken. Do not exceed the specified torque loading of the separator or it may break. A hammer blow on the bolt head should free the joint.

The joint can now be unscrewed. Take care not to disturb the position of the locknut if possible. The track will have to be reset after any ball joint removal anyway, but this will prevent it being incorrect to any large extent.

Where ball joints are badly worn, replacement is the only remedy. It is becoming more common now for manufacturers to supply only complete assemblies, that is the tie rod and two joints, rather than separate joints.

Replace the new joint in the reverse sequence to removal. The tapered pin, however, must be sufficiently engaged to prevent it turning as the securing nut is tightened. This can be done by hitting the top of the joint with a hammer, but often this will not provide the necessary force, and is impractical where space is limited. Also, if there is a grease nipple on the top of the joint, it must be protected by a socket when the joint is hammered.

In some cases, such as joints under the sump, it is impossible to strike the joint. A more satisfactory method is to use a 'G' clamp to squeeze the taper home. Select a suitable socket to fit over the stud of the ball joint, and another if the joint has a grease nipple on top. Tighten the clamp. You will probably have to resort to trial and error in order to gauge the amount by which the clamp must be tightened. This can be a tedious business, but patience is essential. At all times, take extreme care not to damage the rubber dust cover. If this is split, the new joint will be useless in a very short time.

If the joint faces upwards, so that the nut points to the top of the car, it is possible to ease the taper into the eye by applying pressure from below with a hydraulic jack, again taking precautions to avoid damaging any grease nipples.

Finally, tighten the securing nut. After any work on the steering, the car should be carefully road-tested.

25 Unscrew the joint, carefully noting how many turns are needed. Screw in the new joint by the same number of turns

26 To engage the taper of the pin, it may be necessary to squeeze it home using a G clamp and a socket to protect the securing nut

27 With the locknut on the track rod tightened, do up the nut on the joint. Do not forget to use a new split pin

Checking the tyres

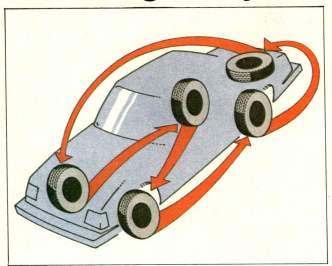

1 Rotating the tyres will save uneven wear so prolonging the life of the tyres

2 Do not rely on the garage forecourt tyre gauge. Buy one for yourself and check your tyre pressures regularly

There is no doubt that many motorists take their tyres rather for granted, giving them attention only when something goes wrong. We hope that these easy maintenance suggestions will help to give a new slant on the value of a properly shod car—properly inflated.

Tyre rotation
Because car suspension systems and types of tyre construction vary, there is no set pattern in which tyres should be switched about in order to equalize tyre wear.

It is sometimes recommended that all tyres, including the spare, should be regularly rotated around the vehicle but check with the handbook as others recommend no rotation at all.

If you are in doubt remember that all major tyre manufacturers have technical service departments who will be pleased to advise you.

If you do rotate your tyres you must not end up with an illegal mixture of radials and crossplies. Do not forget to adjust the air pressures.

Tyre and wheel balance
For smooth running, the front wheels of your car must be correctly balanced. Any vibration through the steering wheel which appears at a particular speed indicates possible imbalance.

Balancing wheels involves attaching small lead weights to the rim of each wheel. This is one job that must be done professionally.

During manufacture, tyres are balanced to prescribed limits, but when a new tyre is fitted to a front wheel, or a wheel that has been balanced is moved to another position, the complete assembly should be re-balanced.

Tyre pressures
Correct inflation pressures are vital to the life of a tyre and to the safe handling of a car and should therefore be checked regularly when the tyres are cold (Tyre pressures normally increase during hot weather or on a long journey as heat expands the air inside, but this increase should be ignored.) Do not forget to replace the valve caps.

Under-inflation is a frequent cause of severe tyre damage which results from an excessive flexing of the tyre. Under-inflation builds up internal heat which eventually weakens the casing and causes rapid and irregular tread wear along the outside edges of the tyre.

Over-inflation also leads to a shortened tyre life, with the tread wear concentrated on the centre section of the tyre. The cushioning effect of the tyre is impaired, resulting in a hard, uncomfortable ride. The tyre is also more susceptible to impact fracture and casing damage.

Different inflation pressures are specified for different makes and models of cars according to loads and speeds, and on some cars different pressures are recommended for front and rear tyres. The correct inflation pressures will be given in your car's handbook, and sometimes they are given on a small plate under the bonnet. Always check the pressures (including the spare) with an accurate gauge.

Tyre inspection
Tyres should be inspected regularly and flints, nails or other objects embedded in the rubber should be removed.

Your inspection may also yield valuable clues to other faults, either in the tyre itself or the cars' steering geometry.

Look for these indications:
Camber wear, where one shoulder of the tyre is more severely worn than the other, is caused by misaligned steering geometry or by high speed cornering. To have the camber of your wheels checked and adjusted, you must take your car to a reputable garage.

Deflation damage, a breakdown of the tyre casing structure which can arise from under-inflation, overloading or running on a flat tyre. A tyre in this condition must be replaced.

Impact fractures, whose symptoms include a casing fracture caused by a severe localized impact. Such damage (not always visible externally) can lead to sudden and dangerous tyre deflation some considerable time after the impact has occurred.

Uneven wear, usually the result of worn bearings, out-of-balance wheels or 'grabbing' by badly adjusted brakes.

Rasped and feathered-edge tread, caused by wheel misalignment (incorrect toe-in or toe-out of the wheels). This requires a precise setting of the front wheels, normally obtainable only from a tyre dealer or a garage.

Acknowledgements

We wish to thank the following for their help in supplying materials and/or labour for photography in this book:

Spark-plug service, pages 12-16:
Champion Sparking Plug Company Ltd,
Technical Services Department,
Feltham,
Middlessex

Points service, pages 18-23:
Robert Bosch Ltd,
Rhodes Way,
Radlett Road,
Watford,
Hertfordshire

Peugeot Automobiles UK Ltd,
(Ducellier distributors)
Peugeot House,
Western Avenue,
Acton, London W4

Carb settings, pages 42-47 and 49-53:
Gower & Lee Ltd,
1 and 2 York Street,
Baker Street,
London NW1

Zenith Carburettor Company Ltd,
Honeypot Lane,
Stanmore,
Middlesex

Drum brakes service, pages 61-65:
Girling Ltd, Parts and Service,
Birmingham Road,
West Bromwich,
West Midlands

Brake hydraulics, pages 80-89:
TOYOTA (G.B.) Limited,
Croydon,
Surrey

Where to buy or hire it

Spark-plug service, pages 12-16:
WD 40 maintenance spray from
Cadulac Chemicals Limited,
Old Boston Trading Estate,
Penny Lane,
Haydock,
St Helens, Merseyside
or hardware stores and accessory shops

Carb settings, pages 42-47:
Colortune kit and Carbalancer air-flow meter:
Gunsons Colorplugs Ltd,
66 Royal Mint Street,
London E1 8LG
and motor accessory stores

Lubrication, pages 54-59:
Coparslip from
The Slip Group of Companies,
Slip Works,
Hatfield Road,
St Albans,
Hertfordshire
or motor accessory shops

Handbrake service, pages 74-79:
Girling Limited,
West Bromwich,
West Midlands

Damper unit fluid, gaskets and cables, all available from Halfords

Brake hydraulics, pages 80-89:
Automatic brake bleed tube from
branches of Halfords

The cooling system, pages 97-103:
Fan belts, thermostats and hoses from
branches of Halfords

Steering checks, pages 112-116:
Ball-joint separator from
Ram Manufacturing Co.,
London
and motor factors

The Hire Shop,
branches throughout Great Britain.
Head Office: Warenne House,
31 London Road,
Reigate,
Surrey

Erratum, page 94
Regarding the adjustment of the clutch linkage on a Ford Cortina Mk III, under the heading 'The exceptions': the clutch pedal should be pumped and then returned to a position level with the brake pedal; not pressed to the floor as stated.

Index

AC Delco distributor 18
air cooled engine 102, 103
  filter 7, 8
Akebono floating caliper
  brake pad 72
  swinging caliper 71
Alfa Romeo engine data 35
antifreeze 101
ATDC mark 26
ATE (Teves) disc brake
  caliper 71
Audi engine data 35
Austin Allegro 69
  Maxi 69
automatic transmission,
  carburettor adjustment 44
  lubrication 54

Ball joints, steering 115
battery lubrication 57
blanking plugs 56
bleeding hydraulic brake
  system 80
  clutch system 93
BLMC 1100 70
  1300 70
  1800 85
  clutch 91
  disc brake calipers 69, 70
  engine data 35, 36
BMW engine data 36
Bosch JFU distributor 18
bonnet hinge and catch
  lubrication 57
boot catch lubrication 57
brake adjusters 61
  adjustment 63
  cylinders 82
  disc 66
  drum removal 62
  dual-line 67, 80
  fluid 67, 69, 80
  hydraulic 80-89
  lights 108
  lining renewal 61
  pad renewal 67-72
  pistons 66-7
  puller 63
  shoe replacement 61, 63
  vacuum servo 80, 89
BTDC mark 28
bubble test, TDC 27

Cable, handbrake 75-9
  lubrication 57
calcium-based grease 56
caliper unit, brake 66
cam followers 31
camber wear, tyres 117
Carbalancer kit 51
carbon contact, distributor 19
  –cored leads 15
carburation adjustment 42
carburettor air-flow meter 43
  adjustment 42
  Colortune kit 42
  compound 46
  emission-control 50, 51, 52
  fast-idle adjustment 43

fixed-jet 43-4, 51
idling screw 44
lubrication 57
multiple 49
progressive 46
sealed 49
twin choke 46
types 43
vacuum gauge 45
variable choke 46
Carter carburettor 43
cartridge-type oil filters 10
Castrol CL grease 56
  LM grease 56
choke, fast-idle adjustment 43
  mechanism 42
  twin 46
Chrysler, Australian
  engine data 41
  gearbox lubrication 54
  Hunter 54
  Imp 55
  lubrication points 55
  see also **Talbot**
circuit tester 106
Citroen brake fluid 67, 80
  engine data 36
clutch, adjusting 90
  bleeding 93
  cable 95
  Girling 93
  hydraulic 90-93
  linkage 93-5
  Lockheed 93
  lubrication 94
  mechanical 93
  pipework 92
  reservoir 91
Colortune carburettor kit 42
  method of use 47
Colt, engine data 36
combustion 25
compound carburettor 46
compression stroke 26
condenser, distributor 19, 23
constant vacuum carburettor 46
contact-breaker points 12, 20-23
cooling system 97
  air 102
  antifreeze 101
  hoses 99, 100
  running repairs 101
  thermostat 99
Coparslip 56, 69
copper-cored leads 16
core plug 100
crossfire, spark-plug leads 16
cross-flow engine 33
cylinders, brake fluid 80
  engine, numbering 25
  front wheel brake 86
  hydraulic brake master 85, 87
  rear wheel brake 83

DAF engine data 37
dashpot 46
  lubrication 57
Datsun 140 J 71

240 Z 71
disc brake calipers 71, 72
engine data 37
Sunny 72
DBA (Bendix) caliper
  brakes 72
diaphragm servo 89
direction indicators 108
disc brakes 66
  calipers 66-78
  fluid 69
  hydraulics 86
  pad renewal 67, 72
  pistons 66-7
  rear 67
distributor cam 20
  cap 18-19
  contact points 20
  HT leads 19
  lubrication 23
  plug lead attachment 16
  rotor arm 18, 19
driveshaft lubrication 58
drum brake wheel cylinders 82
dual-line braking system 67, 80
Ducellier M78 distributor 18
Duckhams MB grease 56

Electrical faults 107
  fuses 106, 107
  lighting system 107
  windscreen washer 110
  wiring 105
electrodes, spark-plug 14, 42
electronic ignition 30
EP oil 54
equalizer handbrake
  linkage 75
  cable replacement 78
exhaust system 103

Fan belt adjustment 102
feeler gauge 14, 21, 32
Ferodo brake pads 67
Fiat 124, 72
  Fiat 125 72
  Fiat 127 32, 72
  caliper brake pads 72
  engine data 37
  valve clearance 32
filters, air 6-8
  oil 8-10
  petrol 10-11
Fish carburettor 43
  adjustment 44

fixed caliper brake pads 66
  ATE (Teves) 71
  dismantling 88
  Girling 69
  Lockheed 70
  Sumitomo 71
fixed-jet carburettors 43-4
floating caliper brake pads 72
Ford brake caliper 69
  carburettor 43, 51
  clutch linkage 94, 95
  Cortina 55, 94, 95
  engine data 37, 38, 41
  Escort 94
  Fiesta 55

V-four engine 33
V-six engine 33
valve adjustment 33
Zephyr 69
Zodiac 69
friction pads, disc brake 66-8
fuses 106

Gap, points 20, 21
  spark-plug 14
  valve/rocker 31
gasket, inlet manifold 42
  rocker box 34
gearbox, automatic 56
Girling adjusters 61
  auto adjust brake 61
  clutch system 93
  fixed caliper 69
  grease 69
  master cylinders 85
  Power-Stop 80
  sliding caliper 69
  swing caliper 69
  tool 70
greasing 56
  brake bearings 69
Gunk degreaser 99
G-valve brake cylinder 85

Handbrake 74-9
  adjustment 75-79
  layouts 74
  lever operation 76
  operation 74
  pad renewal 67
  types of linkage 75
harness, electric 105
headlamps 108
  adjustment 110
  asymmetric 111
heater, air-lock 101
  electric fan 110
  hoses 100
  rear screen 110
Hitachi carburettor 43
  distributor 18
  tamperproof carburettor 52
  twin carburettor 46
HMP grease 56
Holden engine data 41
Holley carburettor 43
Honda engine data 38
horn 110
hose clips 99
  cooling system 98-9
  heater 100
  hydraulic brake 87
  radiator 99
HT leads 15
hydraulic brake, bleeding 80
  disc 86
  drum cylinders 82-3
  failure 80
  hoses 87
  master cylinder 82, 85, 87
  pipes 87
  rear wheel cylinder 83
  vacuum servo 80
  wheel cylinders 82

Idling screw, carburettor 44
ignition, contactless 30

119

electronic 30
light 44
monolithic timing 30
setting timing 27
static setting 28
strobe light setting 28
timing 25
tracking 23
indicator lights 108
inlet manifold gasket 43
interior lights 109

Japanese carburettors 51
Jizer degreaser 99

Kenlowe Kenlomatic
ignition 30
Kiehin carburettor 52
king pins, steering 114, 115

LADA engine data 38
leads, HT 15, 19
insulation faults 15
numbering 19
plug 12, 15
leaf spring lubrication 58
leak seal compounds 100
lighting system 107
linkage, handbrake systems 75-76
mechanical clutch 93-5
steering system 114
lithium based grease 56
line petrol filter 11
load adjuster, headlamps 111
Lockheed brakemaster 80
clutch system 93
fixed caliper 70
heavy duty caliper 71
swinging caliper 70
Lotus engine data 38
lubrication 54-60
Lucas carbon cube 19
dynamo 58
model **23D** distributor 20
model **43D** distributor 19
model **44D** distributor 20

Marelli distributor 19
master cylinder, hydraulic brake 87
Mazda 1500 72
caliper brake pad 72
engine data 39
mechanical clutch system 93
Mercedes disc brakes 72
engine data 39
Metalastik steering column joints 114
MGB brake pad 70
Mikuni carburettor 52
Mini Cooper 70
Mintex brake pads 67
Mitsubishi engine data 36
mixture adjustment 44
monolithic timing 30
Morgan engine data 39
Morris Minor, handbrake cable 77
petrol pump 11
Moskvich engine data 39
multigrade oil 54
multiple carburettor

adjustment 49

Negative-earthed cars 15
Nikki carburettors 43
nine, rule of 32

Ohv engines 31
oil bath air filter 8
changing 9
contamination 8
filters 8-10
grades 54
level checking 54
Opel engine data 39
overhead valves 31

Pad renewal, disc brakes 67-8
paper leaves air filter 7
pedal levelling 93
lubrication 59
petrol filter 10

pumps 10-11
Peugeot 104 111
504 33, 69, 111
disc brake caliper 69
engine data 39
master brake cylinder 85
valve adjustment 33
piston, brake 66-7
clutch 90
servo 89
pitting, rotor arm 20
plug, blanking 56
cap 13
points 14
spanner 14
see also spark-plug
points, contact-breaker 12, 20
gap adjustment 20, 21
Polski-Fiat engine data 39
Porsche engine data 39
positive-earthed car 15
progressive carburettor 46
puller, brake drum 63

Quartz-halogen bulbs 108
radiator 99
rear axle lubrication 54
wheel brake cylinder 83
windscreen heater 110
Reliant engine data 39
removing spark-plugs 13
Renault 12 78
engine data 39, 40
handbrake cable 78
headlamp load adjuster 111
reservoir, brake fluid 82
clutch 91
road testing, hydraulic brake system 82
ignition 30
Rochester carburettor 43
rocker, air filter 7
arm 31
beam action 31
box removal 32
clearance gap 31
cover replacement 34
rod and roller handbrake linkage 75
adjustment 76

rotor arm 18, 19
Rover, brake pad wear 67
master brake cylinder 85
swinging caliper brake 69
rule of five 33
nine 32
thirteen 32
running repairs, cooling system 101

Saab engine data 40
SAE oil 54
screw-in oil filters 9
sealed carburettors 49
servo, brakes 80, 89
Shell Retinax grease 56
short circuit 105
Siamesed valves 32
sidelamps 107
Skoda engine data 40
sliding caliper brake pad 66
Girling 69
Tokico 72
Solex carburettor 43
air screw 46
continental 51
spark-plug 12-16
speedometer cable 59
SPQR valve adjuster 34
static ignition timing 27
steering alignment 113
ball joints 115
box lubrication 59
column joints 114
king pins 114, 115
linkage 114
rack 112, 114
synchronization 49
stop lights 108
strobe light timing setting 29
Stromberg carburettors 43
adjustment 46
emission-control 52
lubrication 57
tools 50
SU carburettors 43
adjustment 45, 46
emission-control 53
lubrication 57
twin 46
Sumitomo fixed caliper brake 71
swinging caliper brake pad 66
Akebono 71
Girling 69
Lockheed 70

Talbot, Avenger 33
engine data 36
firing order 33

tamperproof carburettors 49, 50
taper-seat spark-plug 13
tappets 31
TDC 25
bubble test 27
card 27
mark 26
valve adjustment 33
thermostat 99
thirteen, rule of 32

throttle, carburettor spindle 42
mechanism 42
stop screw 44
timing, ignition 25
marks 26
monolithic 30
setting 27
static setting 28
strobe light timing 29
Tokico sliding caliper brake 72
top dead centre (TDC) 25-7
Toyota engine data 40
tracking, ignition 23
transmission, automatic 44, 54
sealed 54
Triumph 2000 70
twin cable handbrake linkage 75-7
choke carburettors 46
tyres 117
steering faults 113
wear 113, 117

V engine valve adjustment 33
vacuum gauge 45
servo, brakes 80, 89
valve, alternating 32
clearance adjustment 31
feeler gauge 33
numbering 32
Siamesed 32
SPQR adjuster 34
vapour lock brake failure 80
variable-jet carburettor 46, 52
−vacuum carburettor 43-4
Vauxhall clutch linkage 94
engine data 40
valve clearance 32
Viva 94
vernier scale 28
Volkswagen air-cooling system 102
Beetle 102
disc brakes 72
engine data 40, 41
fan belt adjustment 102
volume control screw, carburettor 44
Volvo engine data 41

Water cooling systems 97-101
pump lubrication 59
WD 40 15, 57, 69

Weber carburettors 43, 51
wheel balance 117
bearing lubrication 59
windscreen wiper, electric motor 110
lubrication 59
wire-mesh air filter 8
wiring 105
fault tracing 107
grades 106
replacement 106

Zenith carburettors 43
adjustment tool 50